P

THE EV

Frank Rhodes was ... University of Birmingham. After a year as a ... Doctoral Research Fellow at the University of Illinois he became Lecturer in Geology at the University of Durham, but subsequently returned to Illinois as Associate Professor of Geology. He spent twelve years as Professor and Head of the Geology Department, at the University of Wales, and is at present Vice President for Academic Affairs and Professor of Geology at the University of Michigan. He was awarded the Bigsby Medal of the Geological Society of London in 1967. He delivered the Gurley Lectures at Cornell University in 1960 and the Bownocker Lectures at Ohio State University in 1966. He has served as the Official Visitor of the Committee of Vice Chancellors of Australian Universities, and has lectured in various universities in Asia and Europe under the auspices of the British Council.

Amongst his other books are *Fossils: an introduction to prehistoric life*, *Geology*, and *Evolution*.

His wife is a graduate of the University of Illinois, and they have four daughters.

F. H. T. RHODES

THE EVOLUTION OF LIFE

SECOND EDITION

PENGUIN BOOKS

Penguin Books Ltd, Harmondsworth, Middlesex, England
Penguin Books Inc., 7110 Ambassador Road, Baltimore, Maryland 21207, U.S.A.
Penguin Books Australia Ltd, Ringwood, Victoria, Australia
Penguin Books Canada Ltd, 41 Steelcase Road West, Markham, Ontario, Canada
Penguin Books (N.Z.) Ltd, 182–190 Wairau Road, Auckland 10, New Zealand

First published 1962
Reprinted 1963, 1965
Second edition 1976

Copyright © F. H. T. Rhodes, 1962, 1976

Made and printed in Great Britain
by Cox & Wyman Ltd,
London, Reading and Fakenham
Set in Monotype Imprint

This book is sold subject to the condition
that it shall not, by way of trade or otherwise
be lent, re-sold, hired out, or otherwise circulated
without the publisher's prior consent in any form of
binding or cover other than that in which it is
published and without a similar condition
including this condition being imposed
on the subsequent purchaser.

FOR MY FATHER
AND
IN MEMORY OF MY MOTHER

CONTENTS

LIST OF PLATES — 9
LIST OF TEXT FIGURES — 11
PREFACE — 15

1. Living Things — 17
   The nature of life – the abundance and diversity of life – the patterns of life – the distribution of life

2. Fossils – the Life of the Past — 34
   The nature of fossils – the process of fossilization – the occurrence of fossils – the study of fossils – some fossil case histories – the value of fossil studies

3. Life and Time — 64
   Geological time – time in years

4. The Emergence of Life — 77
   The barren past – Pre-Cambrian life – the biggest gap – the origin of life

5. The Hey-day of Marine Invertebrates — 98
   Creatures of the Cambrian seas – life of the Middle Palaeozoic seas – life in the Upper Palaeozoic seas – land, lakes, and rivers

6. Ancient Plants — 144
   The coal swamps – climate and distribution

7. The Rise of the Vertebrates — 158
   Vertebrate origins – life in the water – life on land and water – early amphibia – life on the land

8. The End of an Era — 188

9. The Dominance of the Reptiles — 191
   The land – the seas – the air – the decline of the reptiles

10. The Air — 211
    The insect vanguard – flying reptiles – birds – flying mammals

CONTENTS

11. Modern Plants ... 221
    Flowering plants – Cenozoic plants

12. The Rise of the Mammals ... 229
    Early mammals – placental mammals

13. The Age of Mammals ... 234
    Archaic mammals – the arrival of modern mammals – South American mammals – the return to the seas

14. The Advent of Man and His Kind ... 252
    Monkeys, apes, and men – ancient man – Neanderthal Man – modern man

15. The Teeming Seas ... 269

16. The Development of Life ... 277
    The expansion of life – the continuity of life – the interrelationships of life – the environment of life – the persistence of life – the evolutionary process

Epilogue ... 306

SUGGESTIONS FOR FURTHER READING ... 307

GLOSSARY ... 310

INDEX ... 323

LIST OF PLATES

(Acknowledgement is made to the authors and publishers in brackets for permission to reproduce plates.)

1(a) Fossil tree-stumps (*Geological Survey photograph. Crown copyright. Controller, H.M. Stationery Office*)
 (b) Track of dinosaur (*American Museum Natural History. Photograph by R. T. Bird*)
 (c) Skin of dinosaur (*American Museum Natural History*)
2(a) Ichthyosaur preserved as carbonaceous film (*American Museum Natural History*)
 (b) *Portheus*, a Cretaceous bony fish (*American Museum Natural History*)
 (c) *Glenobotrydion aenigmatis* Schopf (*J. W. Schopf, Paleobiological Laboratory, University of California, Los Angeles*)
3(a) Nest of six dinosaur eggs (*American Museum Natural History*)
 (b) Reconstruction of the 'Burgess Shale' (*L. B. Kellum and I. G. Reimann, Museum of Paleontology, University of Michigan. Prepared by George Marchand*)
 (c) *Spriggina floundersi* Glaessner (*Professor M. F. Glaessner, University of Adelaide, Australia*)
4(a) *Apatosaurus*, a Jurassic dinosaur (*Painting by Charles R. Knight, Chicago Natural History Museum*)
 (b) Jurassic marine reptiles (*Painting by Charles R. Knight, Chicago Natural History Museum*)
5 Cretaceous scene, with dinosaurs and angiosperm plants (*Mural by Rudolph Zallinger in the Peabody Museum, Yale University, New Haven, Conn. Illustrated in* The World We Live In, *by Staff of* Life *and Lincoln Barnett, 1956, pp. 102–3*)
6 The primitive hoofed mammal *Ectoconus* (*Drawn by John Germann. Courtesy of the American Museum Natural History*)
7 Miocene prairie in North America (*Mural by Rudolph Zallinger in the Peabody Museum, Yale University, New Haven, Conn. Illustrated in 'The Great Age of Dinosaurs'*

LIST OF PLATES

from The World We Live In *by Staff of* Life *and Lincoln Barnett, 1956, pp. 102–3)*

8 Scene at the dawn of the Ice Age (*Mural by Rudolph Zallinger in the Peabody Museum, Yale University, New Haven, Conn. Illustrated in* The World We Live In *by Staff of* Life *and Lincoln Barnett, 1956, pp. 120–21*)

LIST OF TEXT FIGURES

(Acknowledgement is made to the authors and publishers in brackets for permission to redraw figures.)

1 Diagram showing the basis of the Linnean system of classification — 22
2 The use of oxygen isotope studies of fossil shells in the reconstruction of ancient temperatures (*Lowenstam, H. A., and Epstein, A.*, Journ. Geology (*University of Chicago Press*), 62, No. 3 (1954), p. 228) — 59
3 The use of microfossil foraminifera in the correlation of Jurassic age strata in Montana (*Lalicker, C. G.*, Pal. Contr. Protozoa (*University of Kansas*), article 2 (1950), pp. 3–20) — 61
4 The use of fossils in rock correlation (*Moore, R. C., Lalicker, C. G., and Fischer, A. G.*, Invertebrate Fossils, New York (*McGraw Hill Book Co.*), 1952, figs. 1–3) — 68
5 The geologic time scale — 70
6 Typical Lower Cambrian fossils — 80
7 Sponges – living and fossil — 100
8 Coelenterates — 102
9 Molluscs — 106
10 'Worms' — 108
11 Echinoderms — 110
12 Brachiopods — 114
13 Arthropods — 116
14 Trilobites (*Hupé, P.*, Traité de Paléontologie, Paris (*Masson & Cie*), ed. *J. Piveteau*, vol. 3 (1953), fig. 32) — 123
15 Protozoans — 126
16 Bryozoans — 130
17 Graptolites — 137
18 Carboniferous arthropods — 141
19 Palaeozoic plants — 148
20 Coal Measure plants — 154
21 Primitive chordates (*Modified from Simpson, G. G., Pittendrigh, C. S., and Tiffany, L. H.*, Life, New York (*Harcourt Brace*), 1957, figs. 22–24 and 22–23) — 160

LIST OF TEXT FIGURES

22	Palaeozoic jawless fish (Agnatha)	164
23	Variation and development of head shields in cephalaspid ostracoderms (*Gregory, W. K.*, Evolution Emerging, *New York (Macmillan & Co.), 1951, fig. 24-2*)	166
24	Placoderms	168
25	Palaeozoic sharks	170
26	The evolution of the fish (*Colbert, E. H.*, Evolution of the Vertebrates, *New York (J. Wiley & Son), 1955, fig. 39*)	173
27	The origin of the amphibia	176
28	The evolution of the amphibia (*Colbert, E. H.*, Evolution of the Vertebrates, *fig. 93*)	181
29	*Eryops* and *Seymouria*, Permian amphibians	183
30	Diagram of the embryo of higher vertebrates (*Roemer, A. S.*, Man and the Vertebrates, *Chicago, Illinois (University of Chicago Press), 1946, p. 78*)	185
31	The evolution of the reptiles	192
32	Reptiles (*Wilson, M., in Oakley, K. P., and Muir-Wood, H. M.*, The Succession of Life through Geological Time, *London (British Museum, Natural History), 1956, Plate 9*)	194
33	Variety in dinosaurs	200
34	Duck-billed dinosaurs	202
35	Comparison of wings of pterosaur, bird, and bat	214
36	Flying reptiles and birds	216
37	Cycadeoids	222
38	The evolution of the mammals	232
39	Two ancient hoofed mammals from the Eocene	236
40	The evolution of the horses (*Simpson, G. G.*, Horses, *New York (Oxford University Press), 1951, fig. 13, A.M.N.H.I.C.*)	238
41	The evolution of the horses (*Simpson, G. G.*, Horses, *figs. 24, 28, 31, A.M.N.H.I.C., various sources*)	240
42	The evolution of the horses (*Simpson, G. G.*, Horses, *fig. 16, A.M.N.H.I.C., after Edinger*)	242
43	The evolution of the horses (*Simpson, G. G.*, Horses, *fig. 16, A.M.N.H.I.C., from specimens*)	243
44	*Baluchitherium*, a hornless rhinoceros	244
45	Convergent evolution among North and South American mammals (*Simpson, G. G.*, The Meaning of Evolution, *London (Oxford University Press), 1950, fig. 32*)	247

LIST OF TEXT FIGURES

46 The family tree of the primates (*Roemer, A. S., Man and the Vertebrates, p. 175*) — 253

47 Table showing the division of Pleistocene time and the relative durations of human cultures (*Modified after Ebert, James D., Loewy, Ariel G., Miller, Richard S., and Schneiderman, Howard A., Biology, New York (Holt, Rinehart and Wilson), fig. 25-14*) — 255

48 Fossil man. A group of skulls — 256

49 The development of Palaeolithic cultures (*Oakley, K. P., Man, the Toolmaker, London (British Museum, Natural History), 1949, figs. 16, 17c and d, 18a and b, 24a and b, 26c, d, e, and f*) — 264

50 The development of Upper Palaeolithic cave art (*Oakley, K. P., Man, the Toolmaker, fig. 37*) — 266

51 Mesozoic invertebrates — 271

52 Mesozoic cephalopods — 273

53 Rates of evolution in the lungfish (Dipnoi) — 282

54 Natural selection in Pleistocene cave bears (*Redrawn after Kurten, B. (1958), with the permission of the author*) — 289

55 Evolution in Tertiary foraminifera (*Redrawn and modified from figures by Kennet, J. P. (1963) with the author's permission*) — 293

PREFACE

The purpose of this book is to provide a readable and comprehensive account of the history of living things which is both intelligible to the general reader and useful to the student. Although I have assumed no previous acquaintance with biology or geology, I have attempted neither to 'talk down' to the reader, nor to evade the more difficult and controversial aspects of the subject (such as, for example, the origin of life). With this end in mind, the number of technical terms has been reduced to a minimum, although I have included a sufficient number to satisfy the needs of the serious student. They are all simply explained in the text and are also included in the Glossary.

I have attempted to make the account both comprehensive and balanced, giving equal attention to plants, invertebrates, and vertebrates, but I have also tried to avoid a mere catalogue of ancient wonders and prehistoric monsters, by demonstrating the underlying relationships and processes which have influenced living things. It is only by an understanding of the processes by which life developed that a knowledge of the history of its development becomes meaningful.

I am well aware that such a broad approach is a difficult and controversial one. The general reader is just as suspicious of scientific jargon as the specialist reader is of 'popular accounts', and to attempt to cater to both is inevitably to risk the condemnation of both. Yet I believe the attempt is worthwhile. Life has existed on this tiny planet for more than 2,500,000,000 years, and there is no more majestic theme than that of its diverse history. For we are more than mere spectators in this cavalcade of life: there is a peculiar sense in which all of us, specialist and general reader alike, are concerned with and are heirs of the long evolution of all other living things. Emerson once declared that there is 'properly no history, only biography'. The history of life is a biography – our biography, and that of the whole human race. No one can approach such a topic without a sense of humility and inadequacy.

The illustrations are taken from many different sources, which are acknowledged on pp. 9–13. I am grateful to the various authors

PREFACE

and publishers for their generosity in allowing me to reproduce a number of copyright drawings and plates.

It is a pleasure to record my thanks to my former secretary, Mrs Margaret Davies, who has typed the manuscript and assisted in many other ways. I must also thank Mr Trevor Marchant and Mr Ian Howells for preparing photographic copies of many of the illustrations, and Mr Herbert McKee for his help in checking the typescript and preparing the index.

This book was written at the invitation of Professor Michael Abercrombie, F.R.S. I am most grateful for his help and interest. My good friend Dr H. W. Ball of the British Museum has read the entire manuscript and Dr A. J. Charig has read the sections dealing with vertebrates. Both have made many helpful suggestions for the improvement of the text. I am grateful to Miss Joanna Webb and Mr D. Erasmus for the care and skill with which they have drawn the text figures.

My greatest debt is to my wife, for whose enthusiasm and patience no formal acknowledgement can adequately convey my gratitude.

University College of Swansea F. H. T. R.
15 May 1959

PREFACE TO THE 1974 EDITION

THE reprinting of this book has provided the opportunity to revise certain sections in which new knowledge has made this desirable. These include the chapters dealing with the origin of life, geological time, the advent of man and the evolutionary process. I am grateful to the Geologists' Association for permission to include in Chapter 16 a small part of a paper I read to that Society.

Ann Arbor, Michigan F. H. T. R.
23 July 1973

Chapter 1

LIVING THINGS

THE NATURE OF LIFE

'WHAT is life?' – it is a question that all of us ask sooner or later, a question as old as man himself. It is also a question to which there are many answers, and one that is ultimately basic to the whole of human experience. But there is a peculiar sense in which the question is important in a review of the history of living things. What is this common property of life which they all share?

Even this limited form of the question is surprisingly difficult to answer, and the difficulty seems to arise partly because life is unique and cannot therefore be easily defined by analogy or contrast, and partly because it is too complex to be defined in simple concise terms (the same is true of energy and matter). At present we can only describe life as a series of processes which take place within certain complex levels of organization of matter. The most characteristic of these processes are familiar to us all: growth, movement, reproduction, metabolism, irritability, and so on. The basic 'stuff' of living organisms is generally spoken of as 'protoplasm'. Protoplasm has no 'standard composition' (it varies in its properties between different organisms, and even within a single individual), but it consists of a complex mixture of water, various organic compounds (especially proteins, nucleic acids, fats, and carbohydrates), and a number of salts (sodium chloride, calcium carbonate, etc.).

Such a definition of life is far from adequate and it tends to raise more problems than it solves. We know very little as yet, for example, about the detailed structure and organization of the cell, the 'basic building block' of living things. At present, therefore, we can describe life only in terms of its constitution and manifestation. Some readers may object that this is inadequate or even irreverent – that life is 'something' distinct from both matter and processes; expressed in both, yet not itself amenable to scientific analysis. But, from a purely biological view-point, it seems that life can be properly described in terms of 'natural' processes occurring within characteristic organizations of matter. This is in no sense to minimize its complexity or diminish its wonder; nor is it to deny that it may have meaning or purpose. But it is to assert that behind the endless variety of living things, there lies a continuous yet almost inconceivably intricate molecular flow of matter and energy, and it is this which is the basis of life.

It is the hidden history of the origin and development of this process which is the most important and obscure part of the history of life, and although we shall later discuss some ways in which it may have arisen, we can at present study it only by analogy and inference. Once the process had developed and had become established, the theme was provided, and the subsequent history of life, spectacular though it is in both proportions and variety, is merely an endless series of variations upon this single theme.

'Life' could conceivably exist as a vast number of different processes, each quite unlike that by which it is in fact maintained. But life as we know it is a unity: for all their diversity, countless millions of living things over thousands of millions of years have shared a common life process, and in this lies at once both the simplicity and the wonder of life.

THE ABUNDANCE AND DIVERSITY OF LIFE

The abundance of individuals

No one who is familiar with living things can fail to be impressed by both their abundance and their variety. The earth, the sky, and the seas literally teem with life, and this abundance is not confined to the more favourable regions such as tropical forests or shallow seas, for even the more inhospitable regions of the earth support rich and varied communities. The abundance of many individual organisms is almost beyond imagination. A recent investigation of the upper one-inch layer of a soil near Washington D.C. showed it to contain more than 1,000,000 macroscopic animals and 2,000,000 macroscopic seeds per acre. A meadow soil of the same latitude proved to contain more than 13,000,000 animals and nearly 34,000,000 seeds per acre. In a California estuary, macroscopic organisms present in the uppermost eighteen inches of sediment numbered more than 3,000,000 individuals per acre. None of these observations took account of the multitude of microscopic forms which were present. Bacteria, for example, abound in many media; a gram of soil may contain several hundred million.

The prodigality of nature is strikingly shown in the number of eggs, produced by many creatures. A single salmon produces as many as 28,000,000 eggs in a season, and a single oyster may 'lay' as many as 100,000,000 in a single season. In neither case, however, do more than a very small fraction of these survive; indeed, if all the eggs of the oyster were to be fertilized and developed, and the offspring multiplied under the same conditions, the great-great-grandchildren would number 66,000,000,000,000,000,000,000,000,000,000,000, and the shells of a generation would make a mountain eight times the size of the earth! The oyster is not, however, by any means unique in this respect. Dodson has calculated that if one hundred starfish of the genus *Pisaster* (half of them females) each produced one million eggs a year (a conservative estimate) and all survived, in only seventeen generations the number of starfish would exceed the estimated number of electrons in the visible universe (10^{100}). Even in animals with a much slower rate of reproduction, full survival of offspring would lead to enormous geometrically progressive increases in the numbers of individuals. Darwin considered that the average female elephant bears about six young during her life span of a hundred years. If the same breeding rate continued in the offspring, and all survived, a single pair of elephants would produce 19,000,000 descendants 750 years later.

Such is the abundance of life: an abundance which is a major factor in the evolution of living things, because, as

Darwin showed, it leads inevitably to 'natural selection' (p. 285).

The diversity of life

The abundance of living things is reflected not only in the enormous numbers of individuals which exist, but also in the diversity represented by the great number of species which have been described. More than 300,000 species of living plants are known (about sixty times as many as at the time of Linnaeus) and about 4,750 new plant species are described each year. Of the total number of species, angiosperms (flowering plants) comprise about 150,000 species, thallophytes (algae, fungae, etc.) about 107,000 species, bryophytes (liverworts, mosses, etc.) about 23,000 species, and pteridophytes (ferns, horsetails, etc.) about 10,000 species.

The number of known species of living animals is much greater than that of plants. A recent estimate by Mayr puts the number at more than 1,120,000, and, if subspecies are included, the number of named forms is more than 2,000,000. New animal species are being described at a rate of about 10,000 a year. When we think of 'animals' we almost instinctively think first of the mammals, the familiar group of generally rather large and conspicuous animals to which we ourselves belong. Only in a rather secondary sense do we consider fish, frogs, lizards, snakes, birds, and so on to be animals. All these groups collectively constitute the vertebrates and, in spite of the fact that to us they 'are animals', the vertebrates as a whole constitute only about five per cent of all known animal species. More than three-quarters of all described animal species are insects, while molluscs (snails, mussels, etc.), other arthropods (crabs, spiders, etc.), chordates (vertebrates, etc.), and protozoans (unicellular organisms) – in decreasing order of numerical importance – constitute the other more common groups. Sponges, coelenterates (corals, etc.), the 'worm-like' phyla, bryozoans ('moss animals'), and echinoderms (starfish, sea urchins, etc.) are less diverse (coelenterates include about 9,000 recognized species, for

LIVING THINGS

example), and the remaining phyla are conspicuously smaller.

Even this staggering diversity does not indicate the full variety of living things, however, for within a species there may still be conspicuous variation. Our own species affords a ready, though not an exceptionally varied, example. The nature and extent of this sub-specific variation is one of the prime concerns of contemporary taxonomy (the science of classification of organisms), for it appears to be a factor of great importance in the development of new species.

All our discussion has so far been concerned only with living organisms. Life has, however, existed on the earth for at least 2,700,000,000 years, and probably very much longer. Simpson has recently suggested that the total number of species which may have existed since the 'dawn of life' is probably of the order of 500,000,000 and this does not seem to be an excessive estimate. On this calculation more than ninety-nine per cent of all the species that once existed have now become extinct. Such numbers mean little except by comparison and analogy. If we accept this estimate of the number of extinct forms, and if each of these species were allocated a space one foot in width, they would form a continuous chain which would encircle the equator almost four times. If all living plants and animals were arranged in a similar chain with a similar allocation of space, they would cover only $\frac{1}{88}$ part of the equator. This would represent a distance of about 284 miles, which is approximately the direct 'flying' distance between Newcastle-upon-Tyne and Southampton, or Rochester and New York.

THE PATTERNS OF LIFE

Introduction – similarities and differences
Whenever we wish to refer to events or objects or ideas, we must use names, and this applies to animals and plants no less than to other things. We read that Adam, the first taxonomist, coined names for the animal world and these 'kinds' of animals are recognized both by the trained taxonomist and by the

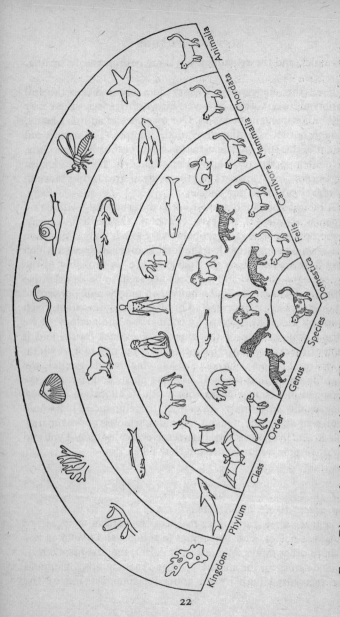

FIG. 1 Diagram showing the basis of the Linnean system of classification, and illustrating the increasing similarity of members of the various taxonomic categories from the Kingdom to the Species.

LIVING THINGS

casual observer. Mayr records the striking fact that the 'primitive Papuan of the mountains of New Guinea recognizes as species exactly the same natural units that are called species by the museum ornithologist'.

By the use of such names any form of organism can be designated. We often find it useful, however, to recognize similarities between different kinds of animals or plants. Thus early man doubtless classified the animals around him as 'dangerous or harmless' and the plants as 'useful or useless' or 'edible or poisonous'. We wander in a garden and naturally refer to weeds, vegetables, fruit trees, flowers, moss, and so on. Aristotle indicated how animals could be characterized according to their habitat, structure, and habits. Later students classified organisms upon their methods of locomotion (e.g. 'creeping things' – worms, molluscs, insects, snakes; 'flying things' – insects, birds, mammals, fishes), their feeding habits (e.g. carnivorous or herbivorous), their reproductive characteristics, their body form, their physiological characteristics, and a host of other features.

Each of these different methods of classification is valid and, within certain obvious limits, useful, and the choice of a particular method or basis of classification will depend largely upon the purpose which it is designed to serve. If the classification is to be of value in the detailed study of living and fossil organisms it must fulfil three requirements. Firstly it must provide a distinctive name for each recognizable 'kind' of organism and the full name must be such as to indicate both the immediate relationships of the organism and also its specific identity. Secondly, such a classification must be readily applicable and must be useful and intelligible to students throughout the world. Local or dialect names are therefore unsuitable, and a standardized system of nomenclature must be developed and enforced by an accepted code. Thirdly, such a classification must indicate genetic relationships. It must therefore combine individual 'kinds' of organisms which are similar, and serve to distinguish them from others from which they differ. To some extent, it must also indicate degrees of relationship. For example, a lion and a tiger are more closely related to each other than either is to a horse, to which, however, they are more closely related than they are to a lobster. These relative differences must be reflected in a satisfactory classification.

The modern biological method of classification

For almost two thousand years after the death of Aristotle (384–322 B.C.) a system of classification based largely upon his concepts was employed by students of living things. The first trend towards the establishment of a more adequate classification was marked by the studies of the English naturalist John Ray (1627–1705) and culminated in the work of the great Swedish botanist Carl von Linné (Carolus Linnaeus 1707–78), whose *Systema Naturae* laid the foundation of our present method of classification. He divided living things into kinds or 'species' on the basis of structural characters, and gave to each species a distinctive name. Linnaeus also proposed a hierarchy of higher categories: genus, order, and class.

The organic world may be divided into two broad groups: the animal kingdom, and the plant kingdom. (Some specialists employ a third kingdom, the Protista, to include the problematical unicellular organisms.) Now the conventional division into two kingdoms clearly includes a very large number of greatly diverse organisms, which resemble one another only in a limited number of characters. Each of the kingdoms may, therefore, be further subdivided into successively smaller groups, each of which will share more characters in common, the ultimate limit being reached with the individual organism.

There are seven main categories of classification and these, together with a number of intermediate forms, constitute the taxonomic hierarchy. The successive restriction of each category may be seen in the illustrated examples (Fig. 1), in which the groups are arranged in decreasing order of size. Some indication is also given to the characters which are used to define groups at various taxonomic levels.

LIVING THINGS

The basic taxonomic unit is the species. Neontologists (students of living organisms) differ widely in their definitions of this category but the following definition is commonly accepted: 'Species are groups of actually (or potentially) interbreeding populations which are reproductively isolated from other such groups.' The problem of recognizing such groups in fossil specimens is complicated both by the incompleteness of the fossil record and by the time factor, which is represented in any sequence of fossils. Some palaeontologists prefer, therefore, to define a species as 'an ancestral-descendant sequence of interbreeding populations evolving independently of others, with its own separate and unitary evolutionary role and tendencies.' A group of closely related species comprises a genus, a group of genera constitutes a family, a group of families an order, a group of orders a class, and a group of classes a phylum.

Linnaeus also devised the system whereby each species is known by a name which includes two words, the first representing the genus and the second the species. These names are sometimes followed by the name of the author who first described the species. Thus the mallard is known as *Anas platyrhyncos* Linnaeus.

One general principle of classification remains to be noted. We have so far tacitly assumed that the 'characters' on which we divide groups are self-evident, but this is scarcely so. 'Why', we may ask, 'are whales classified as mammals rather than as fish?' Admittedly they have warm blood, mammary glands, and lungs – all of which are mammalian characters – but on the other hand they are aquatic, they have fins, they lack hair, and they have fish-like bodies. Why are their mammalian characters regarded as more important in classification than their fish-like characters? From evolutionary studies it has been shown that whales have developed from terrestrial mammalian ancestors, and their fish-like characters are merely adaptations superimposed upon their fundamentally mammalian bodies. The accepted basis for the selection of characters in classification is therefore the phylogeny (the evolutionary history of the race) to which the organisms belong. This is by no means the only possible basis, but it has been agreed by taxonomists to be the most meaningful and useful.

The major groups of animals

Before we can discuss the history of life, it is essential to provide a brief outline of the more important divisions of the animal and plant kingdoms.

One of the most common expressions used to indicate the diversity of the scale of life is the phrase 'amoeba to man'. It is a familiar expression and readily brings to mind the complexity of some animals and the relative simplicity of others. In fact, however, the amoeba is a highly developed organism, which is almost as much more complex than the simplest organisms as man is more complex than amoeba. Below it on the scale of life come many other really simple forms. The flagellates, for example, are minute protozoans (unicellular organisms some of which contain chlorophyll and yet are capable of the rapid movement one normally associates with animals. The Myzophyceae, the blue-green algae, commonly display no separation of the nucleus within the cytoplasm of their cells, and the bacteria, which resemble the blue-green algae in some respects, are even more primitive. Some diseases, however, including measles, influenza, and a number of plant diseases, are produced not by bacteria nor other organisms, but by minute bodies of ultra-microscopic size known as viruses. These consist of nucleic acid and protein, and vary from about 10 millimicrons (ten millionths of a millimetre) to about 200 millimicrons in size. They pass the finest filters but are visible with the aid of an electron microscope, using magnifications of the order of 30,000. Their characters are such that they appear to stand very near the borderline of the world of living things. They bridge the size gap, for example, between molecules and the smallest undoubted living organisms (the smallest bacteria). They resemble living things in their reproduction within the body of an appropriate host, in their existence as definite

LIVING THINGS

strains, each with its own characteristics, and in their susceptibility to changes which are broadly similar to mutations. On the other hand, they can 'exist' only in a living host, and the 'reproduction' may be the result of the reproducing mechanisms of the cell. They have not been shown to undergo any form of respiration, and, most remarkable of all, some of them may be crystallized and stored for a prolonged period, with no apparent loss of their characteristics; for if the crystals are afterwards introduced into the appropriate host, the virus will continue to develop.

These bodies, some composed of only a single molecule, are some of the most problematical structures known. Whether they are living cells, chemical entities, or complexes which assimilate the characteristics of living things from the host which they parasitize, but appear inanimate in the absence of a suitable host, we do not know. Indeed the viruses as a group exhibit sufficient variation in size to make it conceivable that they may include all three. They seem to lead us to the very threshold of life.

Viruses, like almost all the other really 'simple' organisms, are not known as fossils, but they are a useful reminder that the history of life recorded in the fossil record is very incomplete. Below and within our level of observation and description there exist countless other organisms and processes which have played an obscure but vital part in the broader history of life.

The following account and figures deal primarily with those groups which are commonly represented as fossils and, for the sake of simplicity, omit most other groups, even though these are of great interest to students of living organisms. Ten animal phyla make up the great bulk of the fossil record. The main characters of these divisions and their typical representatives are illustrated in Fig. 5. All the major groups are included in the Glossary (p. 310).

The following list gives some indication of the character and content of each of the phyla. A fuller discussion of each is given with the description of its first appearance in the fossil

THE EVOLUTION OF LIFE

record. Page numbers in brackets refer to the appropriate description.

KINGDOM ANIMALIA (*Animals*)

PROTOZOA
: One-celled 'animals', including the foraminifera and the radiolaria (p. 124)

PORIFERA
: Sponges, having porous body walls and commonly a siliceous, calcareous, or spongin skeleton (p. 99)

COELENTERATA
: A single body cavity performs all the vital functions – the corals, jellyfish, sea anemones, etc. (p. 103)

BRYOZOA
: Minute colonial animals with calcareous skeletons – the moss animals (p. 129)

BRACHIOPODA
: Bivalved marine shellfish – the lamp shells (p. 112)

MOLLUSCA
: Highly developed invertebrates, including clams, oysters, snails and slugs, nautilus, squids, etc. (p. 104)

ANNELIDA
: Highly developed segmented worms – including a number of worm-like groups (p. 107)

ARTHROPODA
: Segmented animals with jointed appendages on each segment. Insects, crabs, lobsters, etc. (p. 115)

ECHINODERMATA
: Spiny skinned animals including starfish, sea urchins, sea lilies, etc. (p. 109)

CHORDATA
: Animals with a notochord including all vertebrates (fish, amphibia, reptiles, birds, mammals) (p. 158)

KINGDOM PLANTAE (*Plants*)

THALLOPHYTA
: Simple plants, lacking roots, stems, and leaves, including bacteria, seaweeds, and fungi (p. 144)

BRYOPHYTA
: Plants with leafy stems: mosses and liverworts (p. 149)

TRACHEOPHYTA
: Vascular plants: ferns, trees, flowers, shrubs, grasses (p. 149)

LIVING THINGS

THE DISTRIBUTION OF LIFE

One of the most striking facts that will emerge from our discussion of living things is the way in which their structure adapts them to certain ways of life: the gills of a fish, the eggs of a reptile, the wings of a bird – all these allow their owners to live distinctive types of existence. But the corollary of this adaptation is that the more perfectly any organism is fitted for a particular mode of life, the less capable it is of surviving under different conditions. Pigs don't fly, and whales can't walk! Every living thing therefore has its own peculiar way of life and this is controlled by its structure and physiology and by the nature of its environment. The relationship between an organism and its environment is a delicate and complex one, and its study forms the basis of the science of ecology. Wherever we travel on the earth today, we find living things – from the snows of the highest mountains to the waters of the deepest oceans, and under all kinds of conditions on, above, and within the earth. This great zone of life that envelops the earth is known as the biosphere and it includes three principal media, land, air, and water, within each of which there exist countless smaller but distinctive environments. Thus on the land the animals of deserts are quite distinct from those of the frozen Arctic tundra or the tropical rain forest, and even within each of these environments there will be smaller areas and niches, each with its own individual physical conditions and its own peculiar fauna and flora.

But the teeming life of every nook and cranny of the earth is a thing of the present. Ancient organisms were much more restricted and the slow extension of the biosphere, as one environment after another was colonized, is one of the most dramatic aspects of the long history of life.

Factors controlling the distribution of life
Although it is often easy to observe the general way of life and

broad pattern of distribution of a given species, it is much more difficult to evaluate the various factors which influence its habits and control its range. Nor is this difficulty diminished by the fact that the factors which are of controlling importance in one environment (say a meadow) may be quite different from those which are of importance in another (such as the waters of a lagoon). Now what are these factors? We may think of them as being of three broad types, physical, chemical, and biological. The physical conditions will include such things as temperature, pressure, light and atmospheric conditions, depth of water, viscosity and diffusion of air, water currents, tides, topography, geographical position and composition of land surfaces, physical barriers to migration, conditions of sedimentation, and so on. Chemical factors include salinity, hydrogen ion concentration, gas and organic content of water, the oxygen, nitrogen, and carbon dioxide content of air, and the presence or concentration of a large number of compounds and trace elements. The external biological factors (as opposed to the 'internal' factors represented by the organism's physiology and structure) are represented by the numbers and kinds of associated organisms, their interrelationships as food, prey, competitors, parasites, etc., mobility, population size and rate of change, birth and mortality rates of species, and various other factors.

Few of these factors are constant and most of them exhibit seasonal, diurnal, or random variation, which is often considerable. Clearly, therefore, the equilibrium that exists between organisms and their environment is both dynamic and intricate, and a slight variation in any one factor may have profound effects upon the general stability. There are various ways in which such environmental stimulus and change may affect individual organisms. They may, for example, be driven out of their original niche, they may be destroyed, or they may continue with or without some modification. These modifications themselves may also vary. The greater development of branches and leaves on the sunny side of a tree near a forest margin and the adjustment of the pupil of the eye to light of varying intensity are familiar examples of nongenetic adjustment known as acclimatization. Other genetic changes favoured by environmental pressure may be inherited in successive generations, however, and these are known as adaptations. The process of adaptation is a common one, and is of the highest importance in evolutionary development.

So far we have tacitly assumed that organisms respond as individuals to such changes. This is strictly true, but almost all individuals exist in some kind of association with others, and the response of a group of organisms may be quite different from that of any given individual member. The relationships between individuals naturally vary. A population is any association of one or more species that constitute a closely-knit, interacting system. These interactions involve such things as food competition, parasitic relationships, and so on. Communities are more complex natural groupings, usually involving several populations which inhabit a given area, such as a lake or a forest. All the various factors which influence individual organisms also influence communities: a community undergoes developmental, diurnal, and seasonal changes, for example.

Within such a community individuals may share various relationships. They may, as we have seen, be competitors for food, they may be prey and predator, they may be host and parasite; and each of these broad relationships includes endless degrees of difference. An example will make this clearer. Many organisms live together in intimate mutual association, which is referred to as symbiosis. This may involve the joint association of animals, or the joint association of plants or the association of animals with plants, and the relationship itself may be one of three kinds. Firstly it may be a voluntary association which is mutually beneficial and without damage to either partner. This mutualism is found, for example, in the frequent attachment of sea anemones to the shells of crabs, which provides concealment and the defence of the anemone's 'stinging cells' for the crab and locomotion for the sessile anemone. Commensalism is a voluntary union from which only one of the partners benefits, although the other is unharmed. It may be permanent or temporary, intimate or free living. Small fish and crustaceans often exist in such a way within the tentacles or body cavity of coelenterates, thereby gaining protection and some assistance in food gathering. The third degree of relationship is represented by the involuntary association of parasitism, in which the benefit of the parasite results in progressive damage and sometimes death to the host. Here again, parasitism may be of various kinds, temporary or permanent, sexual (between the male and female of a single species, such as the angler fish, in which the smaller male is permanently attached to the female) or non-sexual.

The distribution of life

The net effect of these multitudes of various factors is to limit the range of species and communities. We have already seen that this limited distribution is reflected in the distinctive populations of different physical environments, such as fresh water lakes, or swamps, or estuaries, and this is one important method of classifying environments.

There are other ways too, however, in which the distribution of living things is reflected. One of the most obvious of these is the presence of broad climatic zones, which vary with both latitude and altitude, and which each support a distinctive fauna or flora. The general sequence of zones is traversed both from the equator to the poles and also from the base to the summit of mountains.

The earth may also be divided into six zoogeographic realms, based primarily upon mammals and birds, although they also support other animals and plants which are more or less characteristic. The Palearctic realm embraces Europe, northern Asia, and North Africa and is characterized by the reindeer, hedgehog, European bison, Marco Polo sheep, and wild ass. North America comprises the Nearctic region, characterized by the caribou, porcupine, pronghorn antelope, American bison, and mountain goat, while South America, the Neotropical realm, supports the giant anteater, sloth, tapir, and

LIVING THINGS

new-world monkeys. The Ethiopian realm includes Africa and the Eastern Mediterranean lands and is inhabited by the zebra, gorilla, giraffe, gnu, aardvark, and African elephant. The Oriental realm of the Far East supports the gibbon, the Malay tapir, the tiger, water buffalo, and Indian elephant, and the Australian realm the marsupial kangaroo, koala bear, wombat, and Tasmanian wolf. These faunas are not everywhere sharply separated, and there is some intergradation, although the realms are generally, though not entirely, separated at present by either oceanic or land barriers (the Sahara desert separates the Ethiopian and the Palearctic realms for example).

Now these differences cannot be explained in terms of different environments. Broadly, similar ecological conditions are found in different realms, South Africa and Australia for example, but the faunas are quite distinct – although in each case the animals tend to assume similar modes of life and even sometimes have a broadly similar appearance. These differences and some similarities are, in fact, explicable only in terms of the geological past. Why, for example, is the fauna of North America more like that of Northern Asia than that of South America, to which land it is now joined by the Isthmus of Panama? The study of fossil faunas from these areas reveals that until relatively recently North America was joined to Eurasia by a northerly connexion, but was separated from South America. These, and other problems, we shall later discuss.

Chapter 2

FOSSILS – THE LIFE OF THE PAST

THE NATURE OF FOSSILS

An interest in the life of the past is almost as old as man himself, as the creation legends of ancient civilizations bear witness. Of man's earliest beliefs we have no records, and we can only guess at the notions which emerged from his growing consciousness. Yet we have at least indirect evidence of his curiosity and acquisitiveness, for in the graves of ancient Neolithic men fossil echinoids have been found carefully arranged around the body. It is paradoxical that man's early belief in a future life should thus be so directly, if unconsciously, linked with the life of the past. But this interest in fossils as curios was by no means confined to early man. The Roman emperor Augustus decorated his villa in Capri with huge fossil bones, which at the time were regarded as the remains of a race of giants. The disc-like fossil crinoid columnals are known from necklaces preserved in an early Christian tomb in northern England (St Cuthbert's beads) and certain tribes of North American Indians long used them as currency. For centuries the Chinese ground up fossil teeth ('dragon-teeth') for medicinal purposes: as a matter of fact the earliest known teeth of one species of fossil man were discovered on the shelves of a Chinese drugstore! In Gloucestershire fossil oysters (*Gryphaea*) were once used in the preparation of medications for cattle, and fossil belemnites for the treatment of eye infections in horses.

But all these are examples of an interest in fossils only as curios and charms – they imply no particular views of, or even interest in, the nature of fossils as such. The oldest records of such views are (like so many other things) provided by a number of Greek writers, some of whom, Xenophanes of Colophon

FOSSILS – THE LIFE OF THE PAST

(c. 614 B.C.) and Herodotus (484–425 B.C.) among them, recognized the true nature of fossils as the remains of once-living things. Aristotle (384–322 B.C.) recorded fossil fishes, and Theophrastus, his pupil, recognized their true affinities, although he attributed their presence in inland areas to fish spawn either left behind in the earth or carried by fishes living in subterranean passages.

Other Greeks, however, not only perceived the true nature of fossils, but also grasped their real significance. Herodotus, for example, realized that the presence of marine bivalves in the mountains of Egypt implied a former submergence of the area. Xenophanes reached similar conclusions about the submergence of Malta, and Xanthus of Sardis (c. 500 B.C.) about Armenia and Phrygia. This may seem to us an obvious and unspectacular conclusion, but it is one which took the world well over two thousand years to assimilate, for it was not until the mid-eighteenth century that there was general acceptance of this interpretation. During the intellectual barrenness of the Dark and Middle Ages other quite different ideas held sway. Fossils were regarded by many as sports of nature; by others as the creations of evil powers, 'intended to mislead or terrify mankind'; by others as unsuccessful attempts by a life force (the *vis plastica* of Avicenna (980–1037) and the *virtus formativa* of Albertus Magnus (1193–1280)) to manufacture living organisms out of the rocks; by others as the products of living seeds carried by wind from the sea or formed under the influence of the stars, and by still others as merely peculiar mineral formations. The widespread and often violent arguments which developed were usually based upon mere speculation, and amidst superstition and dogmatism very little attention was paid to the evidence provided by fossils themselves.

It was not until the fifteenth century that the revival of learning, the development of printing, the re-awakening of interest in exploration, the Reformation, and the founding of universities and learned societies gave new impetus and significance to almost every aspect of human experience and knowledge. No one is more splendidly typical of the new age than Leonardo da Vinci (1452–1519), who during his younger days was employed as an engineer in canal construction in northern Italy. It was here that he observed and studied fossils and correctly interpreted their nature. But such conclusions by no means silenced the controversy over the origin of fossils. Indeed the man who coined the very word 'fossil' – George Bauer (Agricola, 1494–1555) – regarded many (but not all) fossils as inorganic precipitates. A number of beautifully illustrated treatises on fossils were published during the sixteenth and seventeenth centuries, but their authors were no less prone to error than were others. Edward Lhuyd illustrated over a thousand species of British fossils but concluded that they had developed from moist seed-bearing vapours, blown from the seas into the crevices of the earth. Perhaps the epitome of these views is represented by the learned work *Lithographiae Wirceburgensis* published in 1726 by Professor Johannes Bartholomew Beringer, of the University of Würzburg. Beringer accepted the growth of fossils from within the earth and, aided by an enthusiastic group of students, he collected and described a large number of specimens. Many of these are 'good' fossils from rocks of Middle Triassic age, but others are curiously suggestive of carefully manufactured but fabulous monsters, of stars, of moons, and of Hebraic and other letters. It was only after the publication of his book that Beringer discovered that his 'fossils' were elaborate forgeries, carefully carved by his colleague Johann Ignatius Roderique (Professor of Geography, Algebra, and Analysis), aided by three youths, who may well have been students. There have been few stranger episodes of academic intrigue!

By such slow and often painful methods a true appreciation of fossils developed. It was not surprising, however, that amongst those who recognized them as organic remains, many took the further step of identifying them as the remains of the Noachian deluge. Johann Scheuchzer described a fossil skeleton, which he named *Homo diluvii testis*, as the remains 'of one of those infamous men whose sins brought on the world the dire

FOSSILS – THE LIFE OF THE PAST

misfortune of the deluge', but *Homo diluvii* later turned out to be a large fossil salamander.

It was inevitable that the rediscovery of the true nature of fossils should stimulate interest in their significance, just as it had done amongst the Greeks. Both Martin Lister (1638–1712) and Robert Hooke (1635–1703) suggested the possible value of distinctive fossils in the correlation of the strata which contained them, although Lister declared that fossils 'never were any part of an animal'. Hooke went even further and showed how certain Mesozoic fossils from southern England implied not only the oscillation of land and sea but also the existence of a tropical climate in the area, and he suggested that such changes may have been produced by changes in the earth's axis of rotation. These suggestions were neglected for almost a century, however, and it was chiefly the work of William Smith (1769–1838) which established the value of fossils in the recognition of strata in geology. Smith, a surveyor with no formal academic training, put his discovery to good use and in 1815 he published a geological map of England and Wales, consisting of fifteen sheets at a scale of five miles to the inch. This, the first of its kind, was an outstanding landmark in geology and Smith's map has served as a model for all subsequent geological maps.

It was the work of two of Smith's contemporaries, the Frenchmen Lamarck and Cuvier, which provided the foundations of invertebrate and vertebrate palaeontology, and their work, together with Smith's demonstration of the value of fossils, subsequently gave rise to more intensive studies, especially by geologists. These studies received new impetus in the mid-nineteenth century from the publication of Charles Darwin's *Origin of Species*, for in the fossil record alone lay the final court of appeal for evolutionary theory. From then until the present the study of fossils has continued to develop and expand. There are today many hundreds of professional palaeontologists, and a growing and important body of amateurs find both pleasure and profit in the study of the life of the past.

THE EVOLUTION OF LIFE

This study of the life of the past is known as palaeontology (Greek *palaios*, ancient; *onta*, beings; *logos*, discourse) and from what we have already said it will be clear that our knowledge of this ancient life is derived mainly from the study of fossils and the rocks that contain them. Fossils (Latin *fossilis*, something dug up) are the remains of, or direct indication of, life of the geological past, preserved in the rocks of the earth's crust. And so before we look in detail at the life of the past we must turn to fossils themselves. Just how are they formed, and how do we study them?

THE PROCESS OF FOSSILIZATION

Fossilization is a term used to describe a number of more or less distinct processes by which the remains of animals and plants, or evidences of their existence, have been preserved in the rocks of the earth's crust. It follows, therefore, that fossils may be of two kinds. They may, firstly, be the altered or, more rarely, unaltered, remains of organisms, or secondly, they may be direct indications of the former presence of organisms. In one sense such things as coal and petroleum may be regarded as indications of organic activity, but they are not usually thought of as fossils, although they are, however, often spoken of as fossil fuels.

Now animals and plants are not the type of things that are likely to be preserved for any length of time. Not only the soft parts of the bodies but even the hard parts, bones and teeth and so on, decay within a fairly short time, and it is only in very rare cases that organisms are preserved in their original condition. Usually their preservation depends upon the alteration of their perishable parts to more durable materials.

The type of fossilization by which the essentially unaltered remains of an organism are preserved, soft parts and all, requires rapid burial under antiseptic conditions. This is a very rare event, for most organisms decompose rapidly whether they lie on the surface of the land or in the waters of seas and rivers. Occasionally, however, the complete remains of quite ancient animals have been preserved. There are two chief ways

FOSSILS – THE LIFE OF THE PAST

in which this has happened. A few creatures of the last ice age have been preserved under deep-freeze conditions. The best-known of these is the great woolly mammoth (*Elephas primigenius*) which is found in frozen ground in the tundra of eastern Siberia. One specimen of this animal, perhaps 25,000 years old, was discovered by a Russian hunter in 1900 and was so well preserved that the shaggy coat was still present. Part of the flesh had been eaten by wolves, and even the undigested grass in the creature's stomach was preserved. It appears to have met its death by a fall into a crevasse or fissure, for some of the limbs were fractured, the blood was clotted in the chest, and the jaws still grasped uneaten grass. Other specimens of this kind have also been reported, and in gold diggings in certain parts of Alaska fossil mammoth hair is so common in the frozen ground that it often impedes initial mining operations. Another example of the same general type of preservation is that of the woolly rhinoceros (*Rhinoceros tichorhinus*), some soft parts of which were found preserved in asphalt deposits in Galicia, Poland.

Spectacular as these examples are, they are not of great overall importance, both because they are very rare, and because they occur under such impermanent conditions that only geologically relatively recent specimens are preserved. Of greater interest are the more common, though still rather rare, cases in which some of the soft parts of an organism are preserved in more or less altered form. This may be achieved in a number of quite different ways. One of these is the mummifying conditions produced by dehydration. Some of the skin and *faeces* of extinct ground sloths have been found with their bones in the dry caves of Patagonia. Another well-preserved extinct ground sloth skeleton, tendons, and some skin, was found in an old volcanic crater in New Mexico. The 'mummified' dinosaur skin which has been found in various places consists, not of the original skin, but of an impression of it, formed when the dried skin was covered by sand (Plate 1). Not infrequently the impression of soft parts may be preserved as a thin film of carbon which remains behind after the breakdown of the soft tissues and the escape of the more volatile components – oxygen, hydrogen, and nitrogen (Plate 2). Some very ancient fossils are known as remains of this kind. In one deposit in British Columbia impressions of the intestines of arthropods, almost 500 millions years old, are preserved together with numbers of other animals (even jellyfish) which were completely soft-bodied. The impressions of plants which may be collected on any colliery tip are further examples of the partial preservation of soft parts by this process of carbonization or distillation (in which the more volatile constituents are driven off by heat and pressure). The perfect outline of the leaves is preserved in carbon, the more volatile constituents having been lost. It is this same process which has formed the earth's coal resources. In other cases soft parts may be preserved as casts of mud, which filled hollow gut and other cavities.

A well-known example of a somewhat different type of preservation is the case of insects, flowers, and leaves preserved in amber. These have long been known from the Tertiary rocks of the Baltic. The insects became trapped in pine resin, which later hardened around them, and they are preserved as a carbonaceous film which faithfully reproduces even the minutest details of their external structure. These fossils usually, however, consist only of perfect external moulds, although occasionally the muscles and even some of the viscera are preserved.

In some cases the form but not the substance of soft parts may be preserved by rapid burial in volcanic ash. In the great eruption of Mount Vesuvius in A.D. 79, a rain of ash and volcanic debris fell upon Pompeii and Herculaneum, and killed more than 2,000 of the inhabitants. During recent excavations the now solidified rocks have been found to contain hollow moulds of many bodies, of which good replicas can be made. A spectacular example of a similar kind was recently recorded from the basaltic lavas of the Grand Coulee in the State of Washington. The lava yielded an almost perfect mould of an extinct rhinoceros, containing warped and mineralized bones and teeth. The bloated animal appears to have been dead before it was surrounded by hot lava.

The preservation of hard parts

These then are some of the ways in which the soft parts of both animals and plants may be preserved. Most fossils, however, consist only of the hard and more resistant remains of organisms, although the soft parts are frequently represented on them by the various impressions by which they were attached. Here again it is somewhat unusual to find these hard parts in their unaltered original condition. In a few animal groups the chitinous or phosphatic skeletons, teeth, and shells, may be preserved, however, as also may the resistant cuticle of plants. Striking examples are provided by the microscopic fossil jaws of polychaete worms, many of them 500 million years old, which are so perfectly preserved that they are still flexible. Again the teeth of many vertebrates of considerable age and scales of fish are often preserved as fossils with no apparent change in composition. More frequently these hard parts are more or less altered, however, and this alteration usually involves the leaching out of the more soluble components and their partial or total replacement by other substances.

The most simple example of this preservation is represented by shells and bones from relatively young deposits, which usually show evidence of slight but distinctive solution. These may have a rather bleached appearance and all feel light in weight. Most fossils, however, are strikingly heavy, and this is the result of the infilling of the cavities left by decay and solution with other minerals carried in solution. Fossil wood and bone are both rather 'spongy' and are therefore unusually heavy when the pores are filled by this process of permineralization. The mineral deposited in the pores may sometimes be similar to that comprising the original shell or bone, but is usually different, the most common replacing minerals being calcite ($CaCO_3$), silica (SiO_2), and various iron salts. Sometimes this replacement gives a faithful preservation of the original microstructure, as, for example, in much petrified wood, where the cellular structure is often clearly visible in polished sections. Usually, however, although the exterior features of the shell may be unchanged, the detailed structure is lost, either by recrystallization of the original material (the calcite of shells, for example) or by its replacement by another mineral. A small number of fossil eggs have been recorded. The best-known of these were found in wind-deposited Cretaceous sandstones of Mongolia and are those of the dinosaur *Protoceratops*. They are sometimes found in 'nests', and in two eggs the remains of well-developed embryos have been discovered (Plate 3 (*a*)).

The colour of the original hard parts is only very rarely preserved, but there are a few good examples even among very ancient fossils. A fossil brachiopod, at least 350 million years old, shows a series of distinct maroon bands, for example.

Now all the types of fossils we have so far discussed have been the variously preserved remains of organisms. But there are other fossils which are mere ghosts, traces of the former existence of an animal or plant of which nothing any longer remains. An obvious example of this is the leaching away of *all* the original material, so that a hollow mould is left in the rock. This may later be filled in by some other mineral to produce a natural cast of the original organism. These casts and moulds are common in permeable rocks such as sandstones.

Other traces of former life may be rather more indirect, such as fossil tracks, trails, and borings which are sometimes preserved. Some fossil shells display the borings of gastropods and sponges, worm burrows are found in many rocks, some bones of extinct vertebrates show marks made by gnawing rodents, and some tracks, such as those of dinosaurs, are often quite spectacular (Plate 1 (*b*)). Fossil excreta (coprolites) are relatively common and they may provide information about the diet of the creatures they represent. Stomach stones (gastroliths) are sometimes found in association with fossil skeletons. The fossil skeleton of a marine reptile (*Alzadasaurus pembertoni*) found in South Dakota contained within its abdomen 253 of these, which together weighed over eighteen pounds.

FOSSILS – THE LIFE OF THE PAST

Of particular interest are the fossil remains of our own species and of these the most common are the simple stone implements (artifacts), tools, and weapons made by early man. These show a gradual increase in the 'quality' of the workmanship in time and are of great value in determining the relative ages of the more recent strata in which they are found.

These then are the main types of fossils. We have discussed them at some length because they are the foundation of all that will follow, the sole source of our knowledge of the life of the past. But from what we have already said it will be clear that the process of fossilization is a very hit or miss business, and indeed its hazards and vagaries are such that the overwhelming majority of organisms have left no fossil record of their former existence. In most cases the fossilization of an organism depends upon the presence of hard parts (bone, shell, teeth, etc.) and rapid burial. The fossil record is therefore not only very incomplete ('only the skimmings of the pot of life', as Huxley described it), but is also a selective record, lacking almost all trace of soft-bodied organisms and having very few representatives from those environments where rapid burial is unlikely (such as most of those on the land). Even this incompleteness is not all, however, for many organisms which survive all these vicissitudes of fossilization to become fossils are subsequently destroyed by metamorphism (the alteration of rocks by heat and pressure within the earth) or eroded away when the rocks which bear them are exposed to weathering on the surface of the earth.

THE OCCURRENCE OF FOSSILS

We have seen that preservation of an organism almost always involves rapid burial, and there are a number of ways in which this burial is brought about. Some of the more striking ways, such as burial in frozen or oil-soaked ground, in volcanic ash, in resin, and so on, we have already discussed.

THE EVOLUTION OF LIFE

Most fossils, however, are preserved as a result of burial in sediment of one kind or another, which later becomes consolidated and indurated to form sedimentary rocks. Igneous rocks (formed from the solidification of molten material) and metamorphic rocks (those formed from solid, pre-existing rocks which have undergone alteration within the earth's crust as a result of the action of heat and pressure and fluids) rarely contain fossils.

Sedimentary rocks are formed on the earth's surface either by the accumulation and consolidation of rock and other debris, including the remains of any organisms present, or by precipitation from solution. Quite clearly such rocks will be very variable. They will vary in the composition of the fragments of the parent rock, from the weathering of which they have been derived, and they will vary in the kind and degree of weathering which has broken it down. Some will be formed from sediment carried by ice, others from that carried by wind, others still from that carried by running water. Some of these sediments will be deposited on the land: at the foot of mountains, in deserts, at the snout of glaciers, in lakes, in river valleys, in caves, in desert *playas*. Other sediment will be deposited along the coasts: in deltas, lagoons, salt marshes, and estuaries; but most of it will continue beyond these environments and come to rest in some part of the sea, within which again there are many different environments. In each of these many possible sites of deposition there will usually be formed a more or less characteristic type of sedimentary rock, with its own peculiar composition, texture, fabric, and stratification. But even these changes are not the end, for after deposition further variation inevitably arises as a result of quite profound changes within the rock.

These variations in rock type are often reflected in the fossils which such rocks contain. Terrestrial animals and plants will rarely, for example, be found in marine deposits, but are frequently found in rocks which accumulated in lakes or valley bottoms. Similarly deep sea organisms will rarely be present in beach deposits, just as plants confined to the tropics are un-

FOSSILS – THE LIFE OF THE PAST

likely to be found in rocks which were formed under glacial conditions. A shale formed from the mud of a polar sea will contain quite different fossils from a shale of the same age deposited in an inland lake, or a limestone formed in a coral reef. Nor will two rocks formed under identical physical conditions always yield the same fossils, for they may have accumulated at quite different periods of earth history.

Such sedimentary rocks cover an appreciable part of the earth's crust, and wherever they are exposed at the surface, or explored below it, there fossils may be found. Excellent collecting localities are provided by quarries, road and rail cuttings, foundation and drainage excavations, cliffs, canyons and humbler stream banks, and by the host of other natural outcrops that exist. Some of the fossils may be large, others microscopic; some will have weathered out and be obvious at a glance, others will demand careful search and patient extraction; some rocks contain no fossils, others a few, others countless numbers.

The fossils most easily collected are generally those of marine invertebrates, which have weathered out of softer strata. These are abundant in many areas, such as the north-east and parts of the south coast of England, the south-eastern coast of the United States, and so on. Quarries in limestone are often prolific sources of similar fossils and shale from collieries yields many fossil plants.

Fossil vertebrates are much more rarely found by the amateur, and their extraction may often involve quite extensive quarrying. Consider, for example, the work involved in extracting an eighty-foot-long dinosaur, of which only the tail projects from a cliff (and the cliffs always somehow prove to be in the remotest areas). The process calls first for very careful tracing of the position of the skeleton, then the removal of the surrounding rock, and the gradual uncovering of the rock matrix containing the bones. This latter work is usually done with infinite care using a small awl and brush, and each bone must be carefully protected with gum and burlap before being removed. Even the transport presents real problems, for

THE EVOLUTION OF LIFE

a fossilized dinosaur skull may weigh well over three tons, and the other bones are built on the same scale.

In general, of course, this type of fossil collecting is limited, and most fossils are collected without extensive excavation – indeed without any digging at all. In fact one of the most satisfying things about the growth of knowledge of ancient life has been the great contributions which amateur geologists have made to it. The search is open to all, and the only equipment required is a hammer and chisel. It is useful, but by no means essential, to know something about the age of the rocks in any particular area, and to that end the maps and introductory pamphlets listed on p. 282 will be of value.

THE STUDY OF FOSSILS

There was a time when the collection and study of fossils was looked upon as a useless, though harmless, occupation for the man of limited intelligence with unlimited time at his disposal, and the crank. Fossils were regarded as proper components of the dusty depths of a museum, as pleasant curios for use as occasional paperweights – but things far removed from the real business of life – of little, if any, consequence in the broad web of knowledge. There is still a sense in which fossils can be just that: a sense in which their study can be nothing more than a rather superficial commentary on prehistoric monsters and lost creations. But, though this may once have been an adequate description of the status of palaeontology, it is no longer adequate today, for the application of new methods of analysis has revolutionized the study of fossils. What kind of information can fossils provide? How are they studied? These are the questions to which we must now turn.

Fossils are studied by different people for quite different reasons. The palaeontologist working on a well drilled by an oil company is usually interested only in using the fossils to determine the age of the rock from which they come. Another palaeontologist employed in the research laboratories of the same company may be interested in fossils chiefly as indicators

FOSSILS – THE LIFE OF THE PAST

of the kind of 'fossil' environment in which petroleum is likely to be formed. A museum curator may look at fossils only as potential display pieces (though the best curators don't), or again the anatomist may study fossils to trace the evolution of a particular type or character, and so on. But supposing we add all these interests together and ask how fossils may be studied in such a way as to provide every possible scrap of information about ancient life, its character, its environments, and its development. If this is the kind of information we seek, there are four chief aspects of fossils which must each be studied in detail.

Firstly there is the laborious collecting of fossils in the field and after this their extraction and preparation for study. A great variety of both simple and quite complicated chemical and physical methods may be used in this preparation, and it may be a very lengthy process. The extraction, cleaning, repairing, impregnating, and mounting of the skeleton of a large fossil vertebrate, for example, may occupy a team of skilled preparators for three or four years. But here again, this is an exceptional case, and the preparation of most fossils requires much less time.

One of the factors to which the palaeontologist will have paid careful attention in the field is the adequacy of the fossil sample which he collects, for scientific fossil collecting involves rather more than banging an odd shell off a rock face and then walking on to the next outcrop. In an exhaustive study the collector, who must be a trained geologist, will examine the rocks with minute care and will record every band that yields fossils. Some such bands may be a few millimetres in thickness, others may occupy a whole cliff face. From each of these fossiliferous bands he will then collect fossils, and if the band is, say, two metres in thickness he may even collect every centimetre or so. He will pay particular attention to the position and orientation of the fossils within the rock, and probably make accurate measurements of both. He will also make preliminary records of the relative abundance of certain fossils at different horizons, as well as a careful study of the gross characteristics of the rocks themselves. Careful collecting of this kind, though rare, is essential if fossils are to provide all their information. Having probably spent several days collecting from an outcrop only a few feet in thickness, the palaeontologist will then trace the outcrop laterally, perhaps a few hundred yards, perhaps many miles, depending upon the scope of his study, and then begin the process all over again. It is important to realize the detailed and meticulous nature of this field work, for all the study that follows rests upon it.

Having collected and prepared his fossil samples for study, the palaeontologist is now in a position to examine the sample itself. Most fossil assemblages contain quite a large number of different types of animals or plants, and the first step is to make a rough analysis of the various types that are present. There may, for example, be vertebrates and invertebrates and amongst the invertebrates there may be, let us say, arthropods, molluscs, worms, and brachiopods. One of the first things, therefore, which the palaeontologist will want to know is the relationships that once existed between each of these once living groups, and his first examination of the fauna as a whole may tell him something about these relationships. Having completed this sorting of the fossils into different groups he is now ready to undertake the measurement and analysis of each of the various kinds of fossil. On each specimen he will often make three to four or perhaps ten or twenty sets of measurements and as each sample may contain as many as a hundred individuals of any particular species, this is another slow and painstaking job. Partly on the basis of these measurements it is now possible to make a number of very important predictions about the population as a whole. The palaeontologist is able, for example, to tell whether or not the population is a homogeneous one, whether it contains within its numbers relatively comparable numbers of young, mature, and aged forms. He is able to learn something of the range of variation displayed by members of the group, and furthermore, he is able to make statistical predictions as to whether or not his sample is truly representative of the population from which it

FOSSILS — THE LIFE OF THE PAST

was extracted. He may find on examining the results of his measurements that sorting by waves and currents has limited his sample to one particular size of fossil and he will then find himself in a difficult position so far as conclusions about the population are concerned.

When the study of one particular group of fossils is concluded the palaeontologist will turn to other components of the sample in which he is interested, and this again will be a long but important job. Having examined each of the fossil groups the palaeontologist will now compare one group with another and here again interesting relationships may develop. He may find, for example, that one group occurs in direct proportion to another and he may, from a comparison of these relative proportions, be able to make some prediction as to the interrelationship of the groups in the living population of which they once formed a part. He will then be faced with one of his most difficult problems, for he must use his judgement to decide just how closely this fossil assemblage (a thanatocoenose) corresponds to the once living assemblage (biocoenose), for the two are never the same. Similarly, he may find that in the rock itself one group always occurs in a constant position relative to another. One small fossil gastropod, for example, is often found attached to the top of a fossil crinoid, and this suggests that during life there was a symbiotic relationship between the two, in which the gastropod fed on the material excreted from the calyx of the crinoid. In this study it is always important to make as close comparisons as possible with similar living forms, although frequently of course this is impossible, for many fossils represent organisms which are now extinct.

The palaeontologist's next task is to make a detailed examination of the position and type of preservation of the various groups of fossils. He may find, for example, that although he is dealing with fossil shells which consist of two parts or two valves, one valve is much more commonly preserved than the other. He may find that one valve always occurs in a particular position in the rock or he may find that both valves occur

THE EVOLUTION OF LIFE

together, situated in such a way in the rock that they suggest the fossils have preserved the original living position of the animal. If this is so, important conclusions may be drawn.

From this point, the palaeontologist may now proceed to chemical and mineralogical and isotopic studies of the fossils themselves. The chemical composition of the fossils, the ratio of certain isotopes present within them, and the general type of preservation may all yield information of the greatest importance. It is, for example, possible to obtain an indication of the age of many fairly recent fossils from the ratio of the carbon isotopes preserved within them. Furthermore, as we shall later see, it is even possible to make predictions as to the temperature of the seas in which a number of ancient fossils lived, from the ratio between the oxygen isotopes which are present in their skeletons. These and other studies are just now beginning to contribute information of the greatest importance in palaeontology, and they will undoubtedly increase in importance in the future. Having completed these and other studies of the fossils themselves the palaeontologist will now be in a position to gather all the information together in order to make a preliminary synthesis. He will then proceed to the second major study, the study of functional data associated with the fossils.

This study of functional data is a rather more direct and obvious study than that which we have just been considering. Such things as burrows, trails, borings, wear or damage on teeth and tusks, and the position of one part of the skeleton in relation to another will almost always reveal something about the animals which they represent. Some dinosaurs, for example, are known only from the trails which they left behind them. The boring of one fossil animal into another again may tell us something about the interrelationships of the two in life. One particularly good example is the relationship in which two fossil fish were discovered. One fish, the larger of the two, had its mouth wide open and within the mouth there was firmly wedged the body of the second smaller fish. Another example is provided by the oreodont (an extinct fossil mammal) known

FOSSILS – THE LIFE OF THE PAST

as *Merycoidodon culbertsoni* which was found in the Badlands of South Dakota. The particular specimen had within its abdominal region the remains of a pair of unborn twins, which were only partly preserved because they retained the foetal cartilaginous characteristics.

Other examples of the study of functional data, though less spectacular than these which we have just considered, are of no less importance and contribute a vital part to an understanding of the implications of fossils.

The third type of study to which fossils may be subjected is a study of their distribution. This distribution may be of three kinds. Firstly, it is important to know just how a fossil group develops in time. The forms which were ancestral to a particular species may be traced in older rocks. The period of existence of a species in terms either of the thickness of strata in which it occurs or in relationship to other contemporary organisms or to a period of absolute time may be measured. The point in time at which it becomes extinct may also be of importance and there may be connected with it some features of peculiar interest, either in the nature of the rocks themselves or in the characteristics of other animals or plants. Furthermore, an abundance analysis throughout the existence of a particular species is often of the greatest interest, for it may show that its abundance increases or decreases in relation to some other external factors. By such methods as these the chronological distribution of fossil species may be studied and may lead to important conclusions.

Secondly, it is possible to study the ecological distribution of fossils. We may, for example, quite readily discover that certain kinds of marine organisms, like corals, are always found in particular rocks. In fact corals almost always occur in marine strata which are relatively rich in calcium carbonate, especially in limestones. Furthermore, it is often important to notice the relationship between one particular fossil and other components in the fauna of which it is a part, and again to study their interrelationships and their relative abundance. Thirdly, the geographical distribution of particular fossils may be

THE EVOLUTION OF LIFE

studied. A factor of considerable interest in this ecological–geographical study is the detailed comparison between a particular type of rock in which fossils occur and the fossils themselves, and this leads us to a fourth general method of studying fossils.

This method concerns a study of the enclosing rocks, for from these rocks it is possible to learn a great deal about the environment in which a particular fossil or group of fossils lived. We may be able, for example, from the character of the rocks, to make quite far-reaching inferences about their mode of formation and the kind of environment in which they accumulated. It is usually simple to decide whether the rocks are residual (i.e. have accumulated in place), or whether they have been transported. If they have been transported, we can often tell whether or not they have been transported over any considerable distance, and so on. There are various kinds of study which may give rise to inferences of this sort. The grain size of the rocks is studied, together with the vertical changes in the mineralogical and chemical composition, the degree of sorting of the grains which are present and their shape, the colour and the thickness of rocks, their lateral variation, various features on bedding planes and erosion surfaces, together with such things as current marks and ripple marks and slumping in wet sediment. These and many other features of the lithology of sedimentary rocks reveal something of the ancient environment of the fossils which they contain.

It is only when all these detailed studies, each one of them often long and arduous, are added together and synthesized into a whole that fossils will yield all the information which may be derived from them. Most palaeontological studies, of course, are not nearly so detailed as those which we have just outlined. In fact a study of the kind which we have discussed would occupy a team of trained geologists for many months; but it is from such studies that the most important advances in palaeontological knowledge will probably come.

FOSSILS – THE LIFE OF THE PAST

SOME FOSSIL CASE HISTORIES

From what we have already learned it may appear that the study of fossils is a rather serious and dull business. Serious it may or may not be, but it is rarely dull for it combines all the characteristics of fishing, hunting, and a good detective story. We can best illustrate its development and techniques by three or four case histories.

Perhaps one of the most notorious fossils is that of 'Piltdown man'. The first fragment of this creature is said to have been discovered in 1908 by Charles Dawson, a solicitor and keen amateur geologist, near the hamlet of Piltdown, Sussex. After the original discovery Dawson claimed to have continued a careful search of the gravel quarry from which the specimen came until ultimately, after three years' search, he discovered a further skull fragment, and he at once communicated his discovery to the Keeper of Geology at the British Museum. Further careful search yielded other skull fragments and a broken jawbone, together with animal bones and flint implements, and on 18 December 1912 the two men presented their findings to the Geological Society of London. They provided a reconstruction of the skull and suggested that the creature was the long-sought ancestor of man – the 'dawn man' – *Eoanthropus dawsoni*, as they christened him. The most striking thing about the reconstruction, however, was that although the cranial capacity of *Eoanththropus* was only a little less than that of modern man, the creature had a jaw of an ape; and this association made nonsense of man's apparent evolution through the sequence of forms suggested by other human fossils. Inevitably opinions were divided and feelings ran high. Some accepted Piltdown man, and set to work to re-think the history of man's development through him; but many were more sceptical, and claimed that the reconstruction involved fragments of two quite distinct individuals – admittedly closely and remarkably associated in the gravel from which they came – but nevertheless distinct. Furthermore, those who accepted the first view were forced to revise not only the pattern of man's development, but also his age, for animal bones said to have been found in association with Piltdown man suggested an age of about 500,000 years – far more than had ever been guessed before. Three years later Dawson announced the discovery of further skull fragments, together with a lower molar tooth of the kind already observed in the lower jaw. The critics were temporarily silenced, but they remained restless.

Discussion continued intermittently for the next forty years, until a group of three scientists, Professor Sir Wilfrid Le Gros Clark and Dr J. S. Weiner of the Anatomy Department at Oxford, and Dr Kenneth Oakley of the Palaeontology Department of the British Museum, undertook an exhaustive re-examination of the original specimens. A careful anatomical comparison was made of each of the fragments, X-ray photographs were taken, and the nitrogen and fluorine contents of the bones were measured by delicate microchemical techniques in order to provide a guide to their relative ages. In bones preserved under broadly the same conditions the fluorine content increases steadily with increasing length of time of burial, while the nitrogen of their protein is lost at a rather slow and almost uniformly declining rate. The measurement of nitrogen content was therefore used as a supplement to the fluorine studies, and it has the particular advantage of being useful for specimens of too recent age to be amenable to the fluorine method. The surface coloration of the bones was also subjected to searching chemical analysis. The result of the studies is now well known – the scientists concluded that there had been a most elaborate and carefully planned hoax. The jaw was found to be that of a modern ape, which had died at the age of about ten years: it had been artificially stained to match the appearance of the skull and the molars had been carefully abraded to make them appear worn in characteristically flat human fashion. The canine is a young incompletely rooted tooth, but it is, in fact, abraded so much that the pulp cavity is exposed. This is typical only of aged individuals, and explicable only as

FOSSILS – THE LIFE OF THE PAST

the result of artificial abrasion. The brain case was found to be genuine, an Upper Pleistocene variety of modern man, but its origin is unknown. Part of the 'second Piltdown skull' was also shown to have been faked. X-ray spectrographic analysis of the stains of the flint implements, said to have been associated with the bones, showed one of them, a very important one since it was 'discovered' *in situ*, to have also been artificially stained by potassium dichromate.

Piltdown man had guarded his secret well, and many, but by no means all, of the world's leading anthropologists had fallen for the hoax. But the bursting of the Piltdown bubble has a moral of more importance than that which is usually drawn from it (that the most august and dogmatic scientific pronouncements may be based on questionable data). It is a reminder of the new and rapidly developing techniques which are now being brought to bear on the study of fossil remains.

A second fossil, equally well known, perhaps 'the most precious, the most beautiful, and the most interesting' yet discovered, is *Archaeopteryx*, the 'link' between reptiles and birds. Only three specimens are known, all having been found in the famed Solenhofen lithographic limestone deposit of Bavaria; two were discovered in the last century and the third in 1958. One of the specimens is now exhibited in the British Museum, whose first director, Sir Richard Owen, first described it almost a century ago.

The Jurassic rocks of Solenhofen include very pure limestone which is so fine-grained that it has been extensively quarried for use in lithographic printing. This important property, which results from the very quiet conditions under which the rock accumulated in shallow lagoons, protected by fringing coral reefs, has made it of particular interest to geologists, for it has yielded well over four hundred species of fossil animals, in many of which the delicate impression of soft tissue has been exquisitely preserved. No fewer than eight kinds of jellyfish (one of the softest of all soft-bodied organisms) have been recorded, as well as more than a hundred insect species. The preservation of such delicate organisms could only have occurred in conditions of very rapid burial. Other animals include fish, ammonites, pterosaurs (flying reptiles), and crustaceans.

The brightest gem in this priceless collection, however, is *Archaeopteryx* (for details see p. 218). *Archaeopteryx* possessed a considerable number of both distinctively reptilian and distinctively avian (bird-like) characteristics. Its structure was therefore interpreted as a mixture, in which some characters were similar to the class (reptiles) from which it evolved, and others similar to the class (birds) into which it was evolving (see p. 217). Its characters were not intermediate between the two classes, but were a mosaic (as de Beer calls them) with both primitive and specialized components.

This much was known from the original study and description of the specimen, and its great significance as a most remarkable transitional form was realized. But important as the conclusions were, other important questions were unanswered. One of these was of particular importance, for some had used the characteristics of *Archaeopteryx* as an argument against the generally accepted views of the process of evolution (the action of natural selection upon 'normal' variation and random mutations. See p. 285). These critics argued that it was inconceivable that '*Archaeopteryx* had been evolved at random and by the requisite astronomical number of variations, each by itself a hindrance, which gradually transformed arms into wings, provided a sternal keel and girdle, the aeration of bones and skin-sacs, the warm blood, the feathers, and all the essential modification of reptilian structure'. The basis of this criticism is that evolution (as understood by most palaeontologists) demands an intricate and interlocking series of adaptations which are altogether too complex to be brought about by 'random' natural selection, and which are made even more unlikely because intermediate stages in the process of development could not be advantageous.

Now although there are certain inherent fallacies in such criticism, it does serve to illustrate the importance of such 'missing links' as *Archaeopteryx* in an understanding of the evolutionary process. The two slabs of limestone which bear

FOSSILS – THE LIFE OF THE PAST

the remains of *Archaeopteryx* have recently been re-investigated by Sir Gavin de Beer. Before this the sternum (breast-bone) of the creature had been unknown, but with the aid of ultra-violet illumination (which caused the bones to fluoresce) de Beer discovered and dissected the sternum. He was able to show that it lacked a keel – it was therefore reptilian in general character – and that it was of so spongy a texture that it was only poorly ossified and largely cartilaginous. In fact, it required the use of X-ray radiation to demonstrate that the sternum was bony, rather than entirely cartilaginous, in composition. The clear implication is that *Archaeopteryx* lacked strong breast muscles and hence glided, rather than flapped its wings. This conclusion was confirmed by the detailed description of the brain, for de Beer also used ultra-violet photography to show that the small size of the cerebellum was quite inappropriate to control the balanced and complexly co-ordinated movements required by a flying bird.

But, as in the case of Piltdown man, these new studies of an old fossil had wider implications; for the new information about the structure of *Archaeopteryx* showed that it is altogether possible that flying birds evolved from non-flying reptiles, not in one 'big jump' but in a series of steps, which did not require the intricate synchronization of characters which the critics of evolution had supposed.

There was, in fact, no integrated evolution of coordinated 'typical avian' characters in nature. The characters of birds appeared not together, synchronized, and combined, but one by one, slowly, irregularly, piecemeal, and there existed no totality of bird characters (as we define them) until the last had been added, long after the first. To argue as to whether *Archaeopteryx* is a reptile or a bird is to overlook the fact that both categories are but our own abstract concepts, which are neither 'thresholds, shelves, nor barriers' in the continuity of the evolutionary process. The implications of this single fossil specimen are admittedly greater than those of most others, but it is a useful reminder of the unique importance of fossils in our understanding of the evolutionary process.

THE EVOLUTION OF LIFE

One of the most striking examples of the use of isotopic studies of fossils is provided by the work of Urey, Lowenstam, and Epstein, who have shown that the oxygen isotope (O^{18}/O^{16}) ratios of the calcium carbonate of fossil shells may be used to infer the temperatures of the seas in which they lived. It is known that the oxygen isotope ratio of modern shells varies with the temperature of the environment, but before this variation could be applied wholesale to fossils a number of other important factors had to be evaluated. Was the temperature record (the isotope ratio) preserved faithfully through long periods of geological time? Could an organism have physiological mechanisms by which it could secrete shell material which is not in isotopic equilibrium with its surroundings? Are shells secreted only during a limited portion of the temperature range (as in, say, the warmer part of the year) or during the whole of it? And so on. These and other factors were carefully considered and for some of them it proved possible to make appropriate corrections. In spite of other uncertainties which remained it was shown, however, that the study of oxygen isotopes provided a potential method of studying trends in temperature, both chronological and geographical. Studies of belemnites, pelecypods, and brachiopods were made, together with the rock matrix from which they were collected, by means of which it was possible to interpret the temperatures of western Europe and North America during the Upper Cretaceous period (about 90,000,000 years ago). A summary of the results is shown in Fig. 2.

Our fourth 'case history' concerns the study of fossils as a means of correlating rock units. All fossils are more or less useful in this respect, but among the most valuable are those of microscopic size (microfossils). A single small sample of shale may yield many hundreds of perfect specimens, and they have the particular advantage of being obtainable from cores and cuttings from boreholes. Many types of fossils fall within this category, the most important being foraminifera, radiolaria, ostracodes, conodonts, diatoms, algae, spores, and pollen.

Such fossils are useful in both broad correlation of rock units

FOSSILS – THE LIFE OF THE PAST

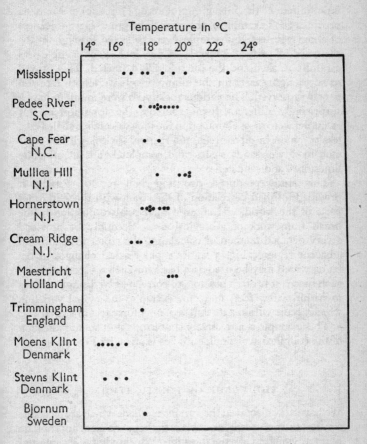

FIG. 2 The use of oxygen isotope studies of fossil shells in the reconstruction of ancient temperatures. Mean temperature distribution of fossil belemnite cephalopod shells from the Cretaceous rocks of various areas. (*After Lowenstam and Epstein*)

THE EVOLUTION OF LIFE

and in more refined and detailed work. Lalicker's study of the foraminifera of the Ellis group of strata (Jurassic) of Montana provides good example of their use. The group has a thickness of about 300 feet, and samples were collected throughout the sequence at vertical intervals of one foot. Each sample was carefully washed and the microfossils extracted. This proved to be a lengthy task, for the minute fossils were both rare and poorly preserved. The various rock types were found to have a surprisingly different fossil content: the foraminifera, for example, were most abundant in the grey calcareous shale beds, less common in olive-green and maroon shales, still less common in silty or sandy shales, and completely absent in oolitic limestones and sandstones.

This study produced two important results. Firstly, it provided a broad correlation of the strata with those in other parts of the world, and showed their relative position in the broad framework of geological time. Secondly, it provided a very detailed method of correlating the strata with those in adjacent regions. Fig. 3 shows rapid vertical changes in the fauna which may be used as a basis for such a correlation. By such means it is often possible to correlate two distant sections to within a few feet; note, for example, the faunal variation which occurs within a thickness of only forty feet of strata.

This example is a relatively simple one, but some indication of the complexity of similar studies is given by Fig. 4.

THE VALUE OF FOSSIL STUDIES

We have now discussed the various methods which are available for the study of fossils, and this is perhaps the most appropriate point at which to ask the question 'Why study fossils anyway? What use is it?' What people generally mean by this question is 'Of what practical use is it?' Suppose first of all we take this question as such – of what practical use is the study of fossils? Mankind down the ages has valued fossils for different reasons, and we have already seen something of the 'fossil

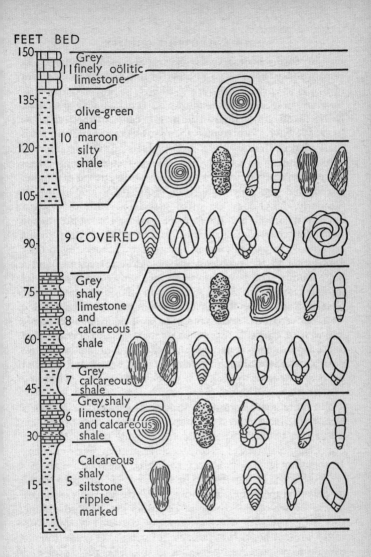

FIG. 3 The use of microfossil foraminifera in the correlation of strata of Jurassic age in Montana. (*After Lalicker.*)

THE EVOLUTION OF LIFE

uses' of fossils as ornaments, charms, and medications. There remain today, however, a number of important practical uses of fossils as such. Fossils are so abundant in some areas that they are used for road-making and other similar purposes. Other fossils are of interest to man because of their value as fuels. We have already noted that petroleum and coal are fossil fuels; that is, they are the direct products of the 'metamorphism' of organic remains. Other less spectacular but nevertheless economically important mineral deposits such as diatomite, guano, and probably some iron ores have a similar origin. Similarly, fossil logs are often associated with the richest deposits of uranium in the form of a mineral carnotite, in the Four Corners region of the western United States. But fossils are equally valuable in a more indirect way, especially in their application to the problems of geological correlation, and a large number of palaeontologists are employed by oil and mining companies in the exploration of sedimentary rocks. Fossils afford by far the most important method of correlating strata in separated areas, and this is of enormous economic importance. Particular examples are the problems of comparing one oil-producing region with another and predicting future possible productions, the tracing of coal seams and mineral deposits, the investigation of bedrock problems in engineering construction, and so on.

But quite apart from these 'practical' uses of fossils, fossils have an intrinsic interest and importance of their own, and this interest is twofold. Firstly, fossils provide one of the most important methods of reconstructing ancient environments. In this they have not only a broad value but they are also, as we have seen, of importance in such detailed studies as those of palaeotemperature. Secondly, they are the only ultimate source by which the course of evolution can be studied, and this is the use with which we shall be particularly concerned in the present book. It is possible, of course, to study minute evolutionary changes by means of modern zoological and botanical techniques, but the chief limitation of these studies lies in the fact that it is impossible to trace animals or plants over a sufficiently long period of time for any major changes to take place within them. In spite of the rather extreme methods (such as exposure to radiation) which are employed to induce various artificial changes, it remains true that they have not yet provided direct evidence of gradual transformation from one species to another. In the fossil record alone there lies the opportunity to examine the whole course of evolution. This is the task to which we shall now turn.

Chapter 3

LIFE AND TIME

GEOLOGICAL TIME

WE have already seen that it is possible to learn a great deal about a group of ancient organisms and the environment in which they lived by means of a study of fossils and the rocks in which they occur. Any such study of a particular stratum will reveal something of this ancient life at one particular 'instant' or limited period of geological time. It therefore follows that by tracing the fossils through the overlying series of strata, we may obtain a more or less complete history of the life of the area at correspondingly later (or more recent) periods of time. In this way the history of life is slowly built up, stratum by stratum, little by little.

But here the geologist faces a major problem, for it is a matter of the greatest importance to determine the mutual time relationships of (to correlate) the fossil sequence of one area to that of another. No single area contains a complete sequence of all fossiliferous strata, deposited and preserved from the first appearance of fossils until the present. A fragment is found here, a fragment there, and the total picture will emerge only as these isolated 'stills' are collected, compared, matched, and spliced into place to form the broad cine film of life. And broad this film certainly is, for each frame is a composite. At any given period of geological time there were many environments, each with its distinctive fauna and flora. Thus at a period of time known as the Jurassic, there were great dinosaurs, tiny mammals, primitive cycads, birds, and insects on the land, ammonites, gastropods, pelecypods, and echinoderms in the shallow seas, and various marine reptiles in the open oceans. Three outcrops may therefore be of similar age – but contain quite different fossils. It is only after each of these environments has been related to others which existed at the

LIFE AND TIME

same period of time and these then compared with others of different age that it becomes possible to build up the composite framework which is known as the geological time scale. This is a standard sequence of events, and the events are of two main kinds, the first kind biological, dealing with changes and developments in animals and plants, and the second kind physical, dealing with such things as growth of new mountain ranges and changing patterns of geography, sedimentation, climate, and so on. We shall return to a discussion of this geological time-scale shortly, but we need to note at present that it is only by correlation and compilation that we can get a complete record of geological events at all.

We have already seen that the aim of correlation is to determine the time-relationship of rocks in one area to those in another, and we have also seen something of the importance and the need for this correlation. But how is it done? How in fact may we take two completely isolated outcrops of rock and determine their correspondence in age? There are many different methods which the geologist employs and these are so varied and some of them so technical that it would be pointless to discuss them in detail. We may, however, summarize briefly some of the simpler ones. In general, in an undisturbed series of strata the oldest is that at the bottom and the youngest that at the top. The first task in correlation is usually therefore to determine the age relationships of the strata in local exposures and then to determine their relationship to strata in other areas and their position in the broad geological time scale. One obvious method of correlation is simply to select the outcrop of a bed of rock and walk the outcrop so as to trace it laterally. This can also be done by means of aerial photographs, especially in arid or semi-arid regions, but this general method is, however, of very limited use, because all beds sooner or later either pinch out laterally or merge into others.

This method of correlation tacitly assumes that the bed of rock in question is everywhere of the same age. This is generally more or less true, although there are important exceptions.

An essentially similar method is that used in tracing a distinctive subsurface bed or sequence of beds in boreholes.

Such general methods of correlation by physical continuity will lead to a more detailed examination of the physical character of rocks, and this provides a second general method of correlation. In this method gross rock types (sandstone, shale, volcanic ash) are matched, and use is made of microscopic features of fabric, sedimentary structures, grain size, sorting, and mineral composition. Some physical and chemical characteristics are particularly useful in petroleum exploration where correlation has frequently to be made in deep boreholes, without the aid of adjacent surface outcrops. These methods include such things as the resistivity and self-potential of a rock (measured by lowering an electrode down a borehole), its radioactive properties, and the time taken to penetrate a unit distance under standardized drilling conditions (drilling time). The most common methods of geophysical reconnaissance include gravity, seismic, and geomagnetic surveys, and a number of other physical and chemical methods of correlation are also being developed. In broader problems of correlation such things as erosion surfaces, the relationships of igneous and metamorphic rocks, and deformation of one kind or other are often employed.

All these methods involve study of the physical properties of the rocks themselves and as such all are limited to more or less restricted areas because of the rapid lateral facies variation (in the character of the rock, its lithology, and its fauna) which all rocks may display. The second type of method of correlation, which depends upon the use of fossils, is rather less limited than the first, for although communities of organisms are also subject to lateral changes, there are some forms which are less dependent upon their environment than others. Fossils therefore represent the most useful method of correlation, and in fact the only reliable method of long distance correlation.

We have already seen that William Smith first formulated the method of fossil correlation towards the close of the eighteenth century. It depends for its value upon the process of organic evolution (which was unknown to Smith, whose methods were strictly empirical). Living things have evolved continuously, and successive periods of time have each been characterized by particular types, levels, and assemblages of living organisms, amongst which changes (with a few important exceptions) appear to have taken place on a world-wide basis.

The value of fossils in correlation naturally depends upon their relative abundance in the particular group of rocks to be studied, and the methods of using them also vary. The most useful fossils are those which have a short time-range and a wide geographical distribution, and swimming and floating marine organisms are therefore generally more useful than benthic (bottom-living) forms. Correlation may sometimes be established on the basis of only a single specimen (a guide fossil), but usually it depends upon detailed and often statistical comparisons of whole fossil assemblages, of the evolutionary development of certain selected groups and the first appearance of particular species in a rock sequence.

The particular methods of correlation used in any given study vary greatly, and demand highly specialized knowledge. Indeed ultimately all the 'methods' we have discussed are aids to, rather than methods of, correlation, for correlation as such is essentially a personal interpretation and can never ultimately be proved or 'read off' by foolproof methods.

We have seen that one of the aims of correlation is to produce a geological time scale which will serve as a standard frame of reference for the description and correlation of various isolated events in the history of life on the earth. In this, the geological time scale bears a close similarity to the time scale of human history. For example, both human history and geological history are subdivided into chapters or eras, which in human history are generally based upon the periods of supremacy of certain peoples. We recognize, for example, the Egyptian era, the Greek era, and the Roman era, and each of these eras may be further subdivided according to the reign of particular

THE EVOLUTION OF LIFE

dynasties. If we take what we might call the Western era and look at its representation in England, we find that it is divisible into various periods, the Norman period, the Plantagenet period, the House of York and the House of Lancaster, the Tudor period, the Stuart period, and so on. (Readers in republican countries must be patient with the use of this monarchical analogy, for the vagaries of politics and political successions do not depend upon the genetic relationship which

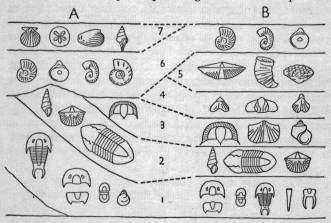

FIG. 4 The use of fossils in rock correlation. The diagrammatic sections A and B show rock successions at two localities, 200 miles apart. (*After Moore, Lalicker, and Fischer.*)

is the essential parallel to historical geology.) Now each of these periods or 'dynasties' that we have just considered may also be subdivided into still smaller units, based upon the reigns of successive kings. Thus the Norman period, for example, may be divided into the reign of William the Conqueror, William Rufus, and so on. In just the same way each reign is divisible into smaller periods of time established upon such events as wars, social changes, legislative acts, and so on. Quite clearly in human history not all the divisions are of equal length or of equal importance, and yet considered as a whole they form

LIFE AND TIME

a connected chain which once established serves as a standard of comparison for all other events within the history of mankind.

It will be noted that no dates have been used, nor in fact has there been a need for them, for by the use of such a 'time' scale (with certain key events arranged in chronological order) it is possible to determine the position of any other event and its *relative* relationship with respect to other events.

It must not be supposed, of course, that the use of absolute dates or absolute time is of no value in the study of human history. It is of great importance to know whether an event took place twenty years after or two hundred years after another event, and in this respect, the scale of time which we have used at present clearly leaves something to be desired. When, however, it is supplemented by a method of dating which involves the measurement of absolute periods of time, it gives a completely adequate framework for the history of mankind.

So far we have considered only human history, but geological history may be divided in just the same way. We may compare the divisions as follows:

Human History	Geologic History Time	Geologic History Strata Deposited	Example
Period of supremacy of a nation	Era	'Rocks'	Palaeozoic
Dynasty	Period	System	Carboniferous
Reign	Epoch	Series	Avonian
Longer periods	Age	Stage	Lower Avonian
Lesser periods	Chron	Zone[1]	*Zaphrentis* zone

The basis of subdivision of geological history is twofold. Firstly, it has been broadly subdivided into eras founded upon the general character of the life they represent. Thus we have the Cryptozoic (hidden life), Palaeozoic (ancient life), Mesozoic (medieval life), and Cenozoic (modern life) Eras. (Many

1. There are many variations of this. See for example Dunbar and Rodgers, *Principles of Stratigraphy* (J. Wiley & Sons), 1957.

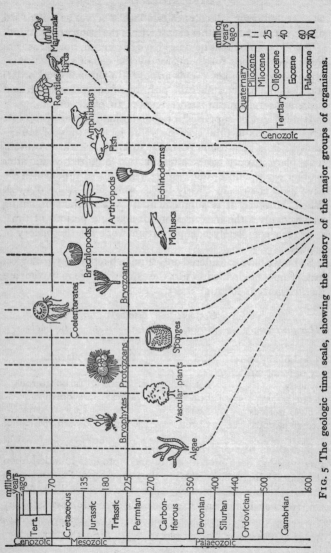

FIG. 5 The geologic time scale, showing the history of the major groups of organisms.

British and continental stratigraphers continue to use an older subdivision of the Cenozoic into two separate eras, the Tertiary and the Quaternary, but there is little justification for this.)

Each of the eras is further divided into a number of periods, based generally upon selected sequences of stratified rocks which display particular fossils. It is, of course, the fossil characteristics rather than the rock characteristics as such that are the basis for the correlation of other (unknown) sequences with these 'type sections'. The names of the periods are coined either from localities where they were first recognized or described (e.g. Cambrian the old Roman name for Wales; Ordovician, an ancient Celtic tribe who inhabited parts of Wales; Silurian, another ancient tribe of the Welsh borderland; Devonian, from Devonshire; Permian, from Perm, a kingdom of old Russia; Jurassic, from the Jura Mountains), or from some distinctive feature of the period (e.g. Carboniferous, the great coal-bearing period of many parts of the world; Triassic, a period represented by a threefold development of rock types in Germany, where it was first described; Cretaceous, the system containing extensive chalk deposits), or, finally, from an older method of classification (e.g. the Tertiary and Quaternary are system names based on an earlier division of geological time in which they were regarded as eras following the Primary and Secondary. This classification has now been generally discarded but the Tertiary and Quaternary persist as period names). It is often convenient to refer to Upper, Middle, and Lower divisions or Early, Middle, or Late portions of a particular period, but the terms are better not employed in strict stratigraphical sense. Smaller subdivisions are based upon local developments and the terminology used varies from place to place.

This calendar of geological time was built up piecemeal, and in many ways it still reflects its origins. It has now been sufficiently well defined, however, to allow its world-wide application, and Fig. 5 shows the general character of its various divisions. A few minutes spent in memorizing the sequence of periods will greatly simplify the reading of all that follows.

One further fact needs comment. Fossils appear in force only in Lower Cambrian times, and this, as we shall see later, is a comparatively late stage of geological time. In Pre-Cambrian times (representing perhaps nine-tenths of all earth history) fossils are at present so rare as to be valueless in correlation, and this must be based upon physical criteria. The result is that present correlations of Pre-Cambrian rocks, even within limited areas, are often imprecise.

TIME IN YEARS

When a palaeontologist is offered a fossil and asked 'What sort of age is this?' he will almost always answer 'It's Jurassic' or 'This is Ordovician' or so on. Now this sort of answer is meaningful as far as it goes and it is one that we frequently use in other cases – we speak of Norman cathedrals and Regency furniture, for example. But what the questioner usually means is 'How many years old is this thing?' In geology as in human history, absolute as distinct from relative time has a particular interest.

Some of the earlier views of the age of the earth are preserved in ancient literature. The ancient Brahmins believed the earth to be eternal, Babylonian astrologers deduced that man appeared half a million years ago, while other peoples accepted much smaller figures for the span of past time. Most early estimates of past time were based upon mere guesses, upon philosophical speculation, or upon some form of intuition. But as long ago as 450 B.C. Herodotus had suggested that the rate of deposition of sediment by the waters of the Nile indicated that the building up of its delta must have occupied many thousands of years. As a matter of fact it was this very method which was amongst the earliest used in the attempt to establish geological time. Geologists suggested that, if the total thickness of sedimentary rocks were determined and then divided by the annual total rate of deposition of sediment, an

LIFE AND TIME

approximate time scale could be obtained. This was done by a number of geologists, and the average estimate of the total thickness of sedimentary rocks was about one hundred miles. But this has proved to be a totally unreliable method, for it is quite impossible to obtain any accurate estimate of the amount of sediment carried into the sea. Even if an approximate figure is accepted, it is quite clear that the rates of erosion and deposition have varied enormously at different periods of earth history and in different parts of the earth. Furthermore many sedimentary rocks are formed from 'reworked' sediments, derived not from the erosion of a parent igneous rock but from an older sedimentary rock, and no adequate allowance can be made for this. Nor indeed can any meaningful allowance be made for either the enormous variations in the total thickness of stratified rocks or the various erosional breaks within them at different places on the earth's surface.

A similar type of method was based upon the sodium content of the oceans. This calculation sprang from the assumption that the oceans were originally 'fresh water' and that all the sodium chloride in existing sea water is derived from the weathering of igneous rocks. The estimated total mass divided by the average annual increment should then provide an estimate of the age of the oceans. This was done and a figure of rather more than 100 million years was obtained. This method of age determination, however, is subject to the same limitations as that which we have just discussed, and it is now known that all the corrections that need to be applied tend greatly to increase the amount of time involved.

Towards the end of the nineteenth century Lord Kelvin suggested that the rate of cooling of both the earth and the sun indicated an age of not more than 20 to 40 million years. The calculation assumed that all the sun's energy resulted from its cooling and shrinking, and, providing this assumption was correct, the time estimate was unquestionable.

Here then were two types of estimate which provided significantly different ages for the earth: the one, apparently incontrovertible, giving a figure of the order 20–40 million years;

the other, suggesting more than twice that maximum figure, yet being subject to major corrections all of which tended substantially to increase it.

The impasse was not finally resolved until the discovery of the radioactive disintegration of uranium by Becquerel in 1896, which proved the existence of a completely unsuspected source of energy for which Kelvin's estimate had made no allowance. But curiously enough, Becquerel's discovery had another, much more important effect upon the measurement of geological time, for it initiated a new field of study which revolutionized concepts in physics and chemistry and ultimately provided a new type of geological clock. The process of radioactive disintegration affects certain elements having unstable atomic nuclei, which undergo spontaneous constant disintegration to form more stable end products. Thus uranium, for example, breaks down to give lead and helium. The rate of this disintegration has been measured with a high degree of accuracy and has proved to be very slow and, like all such disintegration, totally independent of all known environmental conditions, none of the physical and chemical variables which affect most reactions having any apparent effect upon it. Now certain uranium and other radioactive minerals are fairly widely distributed in igneous rocks, although they are almost always found in small amounts. If therefore the ratio of 'disintegrated' lead to 'undisintegrated' uranium is measured, it is possible to calculate the age of the rock. Although there are a number of complications involved in such determinations, they have now provided a reasonably dependable scale of geological time, and they have recently been supplemented by studies of radiogenic lead-thorium, rubidium-strontium, potassium-argon, and carbon isotope ratios.

These successive studies have steadily pushed back the age of the oldest known rocks. The most recent refined strontium experiments show a maximum age of about $3,300 \pm 300$ million years for various lepidolite-bearing pegmatites in Rhodesia, Wyoming, and Manitoba, although other methods of analysis of the same rocks suggest a rather smaller age (2,700

LIFE AND TIME

million years). Preliminary studies suggest still greater ages for other rocks from Swaziland, the Transvaal and elsewhere, perhaps 3,700 million years. Indeed, the Rhodesia – Manitoba pegmatites are themselves part of a granitic suite of rocks that is intruded into a considerable thickness of still older rocks. Meteorites, whatever their composition, all prove to have a common age of about 4,500 million years. Because it is believed that they formed when the earth did, this suggests a similar age for the earth. Lunar rocks range from 3,100 to 4,500 million years old.

The combined studies of physicists, chemists, geologists, and astronomers on meteorites, rocks, and the rate of expansion and present distance of the earth from various outlying galaxies (especially M-31) all tend to suggest an age of about 4,500 million years for the earth itself. There are some who suggest that such a distant date represents not only the time of origin of the earth, but also of an unimaginably greater 'cataclysm' in which the cosmos, and perhaps even matter itself, came into existence. Others, however, interpret the apparent expansion of the universe quite differently, and suggest that it implies a continuous creation of matter. 'May it not be that it is a property of space', Littleton asks, 'that wherever space occurs then matter may appear in it from nowhere, and to just such an extent in total throughout the observable universe as to balance the loss over the frontier horizon of the universe?' Clearly, therefore, whichever of these two hypotheses proves to be correct, the question of the age of the earth inevitably leads us to the fundamental question of the origin of matter itself, and even this, once the most profound and inscrutable of all questions, now approaches the threshold of scientific study. So ancient is the earth on which we dwell, ancient almost beyond our imagining, so profound and as yet so obscure are the problems of its origin, and yet they impinge upon every phase of man's experience and knowledge.

We know nothing of life in other parts of the vastness of the universe. It is not impossible that within its uncharted emptiness, other bodies than ours may support living things. It is

THE EVOLUTION OF LIFE

not impossible that 'life' on other bodies may be utterly unlike anything we can conceive, so unlike it, in fact, that even our use of the word 'life' itself may be quite inappropriate and our description might demand a new vocabulary. But on our own planet there is life and we know something of its history. The earth is about 4,500 million years old, but for the greater part of this enormous period of time there are virtually no fossil remains in fact, undisputed fossils appear in quantity only in Lower Cambrian times, about 600 million years ago, that is, during what may be only the last ninth or tenth of the earth's life (a recent estimate, however, suggests that the time involved may be nearer 740 million years ago). Even this fragment of geological time is almost unimaginably long, however. Suppose that an imaginary tree growing at the rate of $\frac{1}{10}$ inch every thousand years had been planted at the dawn of Cambrian time and had continued to grow ever since. It would now be almost a mile high, more than four times the height of the Empire State Building. A similar tree planted when our own species appeared would be a mere two feet in height, and one planted at the time of Christ only $\frac{1}{5}$ inch.

Suppose we attempt to scale down this history into a calendar in which 1 January represents the origin of the earth and 31 December represents the present day. If we accept a round figure of 5,000 million years as the probable age of the earth each second will represent about 167 years, each minute 10,000 years. The Lower Cambrian would then begin on 18 November. Man would appear at about 11.50 p.m. on 31 December. The whole of recorded human history would fall within the final forty seconds before midnight.

Chapter 4

THE EMERGENCE OF LIFE

THIS book is largely concerned with origins – the origin of man, the origin of the various types of animals and plants, the origin of the environments in which they live, and ultimately the origin of life. But this 'ultimate' applies only to the scope of this book – for beyond the origin of life lie other origins, of compounds, of elements, of matter, and ultimately (really ultimately this time) the origin of origins. There is a sense in which each of these origins is dependent upon its predecessors, and there is a peculiar sense in which the particular problems of the origin of life and matter are both scientific and philosophical, although we shall be exclusively concerned with the scientific.

THE BARREN PAST

There is overwhelming evidence that complex organisms have gradually evolved throughout long periods of time from more simple ancestors, and there is therefore little doubt that the earliest organisms were of a very lowly kind. Given certain conditions, many forms of life leave an adequate fossil record, and the natural beginning to our investigation is therefore to examine the oldest fossils.

Perhaps the most remarkable (and also the most perplexing) thing about the fossil record is its beginning. Fossils first appear in appreciable numbers in rocks of Lower Cambrian age, deposited about 600 million years ago. Rocks of older (Pre-Cambrian) age are almost completely unfossiliferous, although a few traces of ancient organisms have been recorded from them. The difference between the two groups of rocks is every bit as great as this suggests: a palaeontologist may search promising-looking Pre-Cambrian strata for a lifetime and find nothing

THE EVOLUTION OF LIFE

(and many have done just this); but once he rises up into the Cambrian, in come the fossils – a great variety of forms, well-preserved, world-wide in extent, and relatively common. This is the first feature of the oldest common fossils, and it comes as a shock to the evolutionist. For instead of appearing gradually, with demonstrably orderly development and sequence – they come in with what amounts to a geological bang. They are not the oldest fossils – but they are the oldest common fossils. Those which precede them (which we shall discuss shortly) amount to a mere handful of forms in comparison, and show no direct ancestral relationship to them.

Now what are they like, these Lower Cambrian fossils? Are they forms which have some resemblance to living creatures, or are they so distinct that they would be strange and unfamiliar to a modern zoologist? Do they seem to be the 'primordial fauna' or are they more advanced? Are they all much of a muchness – or are they diversified?

There are very nearly 500 species of Lower Cambrian animals described, and this is the second striking fact about them. This does not mean, of course, that there were only 500 species alive in those distant times. Many others must have left no fossil record; others too are undoubtedly as yet undiscovered or undescribed. But the sort of figure represented is important in two respects. First, it clearly implies that by the time we get the first good look at fossils in the Early Cambrian, they are far removed from the simple forms that represented the earliest living things: the earliest common fossils, in fact, throw very little light on the oldest living things. But secondly, even allowing for the various corrections we have discussed, there are relatively few species represented in the Lower Cambrian rocks in comparison with, for example, the one and a half million or so species of animals and plants alive today. On the basis of the number of species we must therefore regard this Lower Cambrian fauna both as diversified in contrast to the earliest (or even early) living things and also as undifferentiated in contrast to later faunas.

But diversity is not reflected only in numbers of species; it

THE EMERGENCE OF LIFE

is also reflected in the number of phyla to which the species may be referred. Here again the Lower Cambrian fauna proves to be unexpected, for it includes no fewer than seven phyla. The most commonly represented are the arthropods (about $\frac{1}{3}$ of the total number of species), brachiopods (about $\frac{1}{4}$ of the total), sponges (about $\frac{1}{6}$), gastropods (about $\frac{1}{10}$), together with a few worms, echinoderms, and coelenterates. In this case also we must look both forward and backward, for the Lower Cambrian is both simple in comparison with its later descendants and complex in comparison with its forebears. If a zoologist, convinced on other evidence of the validity of the theory of evolution, but knowing nothing of the fossil record,[1] were told that fossils appeared in force in the Lower Cambrian, and were then asked to guess their nature, he would almost certainly opt for protozoans – or some more primitive unknown group. In fact, protozoans are one of the few major phyla not certainly represented in the Lower Cambrian, although some possible foraminifera have recently been reported from the Lower Cambrian of Yakutsk, in the U.S.S.R.: the others are bryozoans, chordates, and the higher plants. (Calcareous algae are known, from the Pre-Cambrian.) This may at first seem to demolish the whole theory of evolution. Some students have suggested that it does and even Charles Darwin admitted that the problem of this fauna '. . . may be truly urged as a valid argument against the view [of evolution by natural selection]'. But a number of Pre-Cambrian fossils have been discovered since Darwin's day, and, although they present other problems, they at least confirm the conclusion that the real implication of Lower Cambrian faunas is that the greater part of the major differentiation of animals (but not plants) and all the really big steps involved in the general processes of living material had already taken place in distant Pre-Cambrian times. It is to those rocks that we must turn if we are to discover any traces of the oldest living things.

So much for the backward glance. But the complexity of

[1]. This may sound an impossible combination of characteristics, but such people do, in fact, exist!

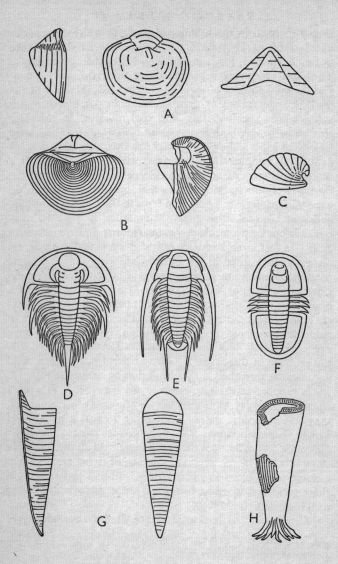

THE EMERGENCE OF LIFE

Lower Cambrian life although real is also relative; in comparison with later faunas, those of the Lower Cambrian are relatively simple. There are no chordates, no fish, no amphibia, no reptiles, no birds, no mammals. There were no familiar plants, no mosses, no ferns, no trees, no flowers (although algae, diatoms, and bacteria were almost certainly present – see p. 83). Most of the invertebrates present had chitinous skeletons and lacked the heavy calcite shells characteristic of so many later forms. Even the phyla that were present were mostly represented by primitive forms: the arthropods were chiefly represented by trilobites, the most primitive members of the phylum; the brachiopods by the modest inarticulate forms; the molluscs by very simple gastropods, and the echinoderms by the long-extinct, ancestral cystoids and edrioasteroids. There are some exceptions to this statement: the worms, the coelenterates, and to some extent the sponges were more complex, although the level of organization of these latter phyla is conspicuously lower than those we have just discussed. But taken as a whole, the Lower Cambrian fauna was simple. It had indeed a relative abundance, diversity, and variety – but only when compared with the longer history of living things of which it was the product.

Lastly, the Lower Cambrian fauna as known at present, though it is world-wide in extent, is exclusively marine. Of life on the land or in inland waters or in the air we have no trace, and it seems certain that such life was a much later development. One of the features of the enormous increase in both the abundance and the variety of post-Cambrian life is that it has taken place by the development of adaptations which have led to the successive invasion and conquest of new environments and new ways of life.

FIG. 6 Typical Lower Cambrian fossils.
A-B Brachiopods: A *Paterina* (natural size, 3 views), B *Kutorgina* (natural size, 2 views); C Gastropod, *Helcionella* (natural size); D-F Trilobites: D *Olenellus* (natural size), E *Bathynotus* (natural size), F *Eodiscus* ($\times 3$); G Pteropod, *Hyolithus* (natural size, 2 views); H Archaeocyathid, *Cambrocyathus* ($\times 3$). (*Partly after Dunbar*)

THE EVOLUTION OF LIFE

PRE-CAMBRIAN LIFE

We have seen that the earliest common fossils are those of the Lower Cambrian, in one sense complicated and diversified, in another simple and lacking great variety. Their appearance in force raises some tantalizing problems, of which the most obvious but the most obscure is 'Where are their ancestors?' Let us now look for them.

The term Pre-Cambrian is applied to rocks which occur below the base of the Cambrian. The base of the Cambrian is defined by the first appearance of a widespread and distinctive trilobite–brachiopod fauna, which also happens to be the first appearance of any substantial mixed fauna in the fossil record. Almost by definition, therefore, we may expect the Pre-Cambrian rocks to be more or less unfossiliferous – and so they are.

Although Pre-Cambrian rocks are usually deeply buried below great thicknesses of more recent sedimentary rocks, there are nevertheless extensive areas where they outcrop at the surface. Sometimes they are exposed in such a way as to reveal their relationships with these overlying rocks: the lowest rocks in the Grand Canyon, for example, are Pre-Cambrian. The two most common types of occurrence, however, are in the cores of great mountain chains, where they have been exposed by uplift and subsequent erosion, and the extensive areas known as 'continental shields', which have remained more or less stable, positive areas in more recent geological times.

Many Pre-Cambrian rocks have been involved in one or several periods of mountain building, and show the effects of intense intrusion, deformation, metamorphism, and subsequent erosion. Many are therefore igneous or metamorphic in character, and not likely to provide any clues in the search for Pre-Cambrian life.

There are, however, great thicknesses of unmetamorphosed and undisturbed shales, limestones, and sandstones within the Pre-Cambrian, and in certain areas (North America, South

THE EMERGENCE OF LIFE

Australia, for example) these underlie the base of the Cambrian without any apparent physical discontinuity. It is in such areas that the most intense search has been carried out, and it is from them that virtually all the accepted Pre-Cambrian fossils have been collected.

Palaeontologists seldom disagree about either the age or the affinities of fossils – still less as to whether a particular object is or is not a fossil. Yet one of the most striking things about Pre-Cambrian fossils is the controversy which almost all of them have provoked concerning their age or their nature or both, and this is itself significant. Of those structures which are generally accepted as genuine fossils, only a very few represent animals. The Pre-Cambrian rocks of the Grand Canyon of the Colorado River in Arizona have yielded an impression that has been claimed to represent a jellyfish, and a number of trails and burrows, at least some of which are apparently of organic origin, have been found in the Beltian Series of Montana, together with some possible (but disputed) remains of inarticulate brachiopods. By far the most important and most convincing of all Pre-Cambrian fossils come from Ediacara, South Australia, and they occur only a few hundred feet below the generally accepted base of the Cambrian rocks there (Plate 3(c)). They include groups of worms, jellyfish, probable pennatulates (sea pens), and echinoderms, as well as some forms of obscure affinities. This Australian fauna, which consists of some 25 species of soft-bodied metazoans, was probably planktonic or pelagic in habit. Similar faunas of comparable age are known from England and South Africa. A number of other Pre-Cambrian animal fossils have been described and dignified by such names as *Eozoon*, *Protoadelaidea*, and *Aitkokania*, but they are almost certainly of inorganic origin, while others are not proved to be Pre-Cambrian in age.

The evidence for Pre-Cambrian plants is rather more convincing, though still somewhat meagre. The oldest of all known organisms consist of bacteria and single-celled blue-green algae and are preserved in rocks in South Africa, that are about 3,200 million years old. Pre-Cambrian Soudan Formation

TABLE 1

Model of mechanisms involved in the evolution of life, composition of the atmosphere and hydrosphere, and geologic data.

For explanation, see text.

(*After various authors, especially Berkener, Marshall, and Cloud.*)

TIME BEFORE PRESENT IN MILLIONS OF YEARS	PROBABLE COMPOSITION OF HYDROSPHERE AND ATMOSPHERE	MODEL FOR DEVELOPMENT OF LIFE	FOSSIL RECORD	GEOLOGIC RECORD
				Cenozoic
				Mesozoic
				Paleozoic
				Pre-Cambrian
	Increase in oxygen			
	Free atmospheric oxygen level adequate for animal respiration	Invasion of shallow marine environments	Cambrian faunas: worldwide Ediacara fauna: Australia, England, South Africa	
	Gradual screening out of lethal UV radiation by build-up of ozone layer		Oldest metazoans, including green algae: Australia	
1,000	Gradual build up of ozone layer	Establishment of respiration by eucaryotic organisms		
	Free oxygen invades atmosphere			
2,000			Blue-green algae and bacteria; Ontario	*Oldest red beds* ←

84

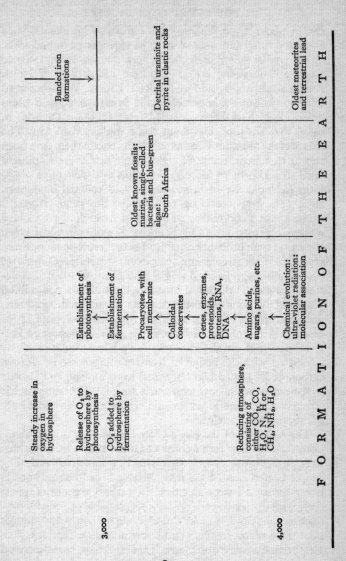

THE EVOLUTION OF LIFE

cherts, about 2,700 million years old, from Minnesota, yield algae and fungi and what seem to be regenerated humic acids. Other similar microfossils, with ages ranging from 1,000 to 2,000 million years, are known from other parts of North America, Australia, and the Soviet Union. Well-preserved micro-organisms, comparable with primitive aquatic blue-green algae, have been recorded from southern Ontario, in rocks that are probably about 2,000 million years old. The cherts of the Gunflint Formation which contain them have been shown to contain traces of eight amino acids. Pre-Cambrian 'coal' from northern Michigan contains oval structures whose form is 'indicative of origin from ovoid colonial blue-green algae of a type occurring today (e.g., the genus *Nostoc*)'. Strata about 1,000 million years old in Australia have yielded metazoan plants, including filamentous and spheroidal green algae that were probably capable of sexual reproduction. From several continents, concentric calcareous structures (*Collenia*) of probable blue-green algal origin have been recorded in limestones, some of which are probably more than 2,000 million years old. Various extensive deposits of anthracite, graphite, and some limestones are regarded by some geologists as indirect evidence of the existence of life, and the important Pre-Cambrian iron ore deposits may have been wholly or partly associated with the action of iron-secreting bacteria. Certain carbonaceous structures (*Corycium*) from Finland and Canada have been shown to have a carbon isotope C^{12}/C^{13} ratio which is strongly suggestive of an organic origin, although some have questioned this conclusion.

The sequence of the earliest fossil organisms is itself significant. The earliest forms consist of unicellular organisms that were probably procaryotic (that is, lacking a nucleus), including simple bacteria, and algae which have a broad resemblance to living forms. All were aquatic. The earliest organisms apparently with nucleated cells (eucaryotes) capable of sexual reproduction are found in rocks almost 2,000 million years younger than these oldest fossils. With a few notable exceptions, Pre-Cambrian animals are unknown. Those that have been described

THE EMERGENCE OF LIFE

(the Ediacaran fauna) are only a little older than the Cambrian, and are wholly soft-bodied. Their affinities, though not always unambiguous, suggest that they are not the direct ancestors of the hard-bodied Early Cambrian fauna, though this is not altogether surprising since they represent a soft-bodied pelagic fauna, while the Early Cambrian is a 'shelly' bottom-living fauna. Some of the Ediacaran fossils, the medusoid jellyfish and the worm-like forms, for example, show a close resemblance to Cambrian forms, but others show little resemblance to later organisms, although one three-rayed form may be ancestral to echinoderms, while other forms may be ancestral to later arthropods.

Most of these discoveries are very recent and have involved the use of elaborate methods of extraction and chemical analysis. They prove beyond all reasonable doubt the existence of primitive, yet organized and well-differentiated, life in early Pre-Cambrian times. They also suggest at least one reason for the otherwise meagre results which have been produced, for the search for Pre-Cambrian life clearly requires not only more extensive and painstaking field studies of promising rocks but also the use of sophisticated techniques in their laboratory examination.

But although the discovery of these fossils solves one major problem, it raises another, for there remains the formidable gap in the record between these forms and those of the Lower Cambrian. To this problem we shall return later.

There is one other indirect source of evidence for the existence of Pre-Cambrian life. The present chemical balance of the earth's atmosphere is largely controlled by various biological processes, and there is fairly general agreement that the existence of atmospheric oxygen and the concentration, though not perhaps the 'original' production, of carbon dioxide are the more or less direct result of biological processes. It has been estimated that the oxygen in the atmosphere is renewed every 2,000 years by the process of photosynthesis, and the atmospheric carbon dioxide every 300 years. Now the geological significance of these two gases is great, for they have a considerable influence

THE EVOLUTION OF LIFE

on both weathering and sedimentation, so that it ought, therefore, to be possible to detect their absence (if they were, in fact, absent) during the formation of older rocks that preceded widespread living things. There is at least some evidence for their absence in such rocks. For example, pyrite, an iron sulphide mineral, and uraninite, a uranium mineral, disintegrate in an oxygen-rich environment, but both are known from detrital pebbles in placer deposits of Pre-Cambrian age in South Africa. It has been suggested that the occurrence of banded, iron-rich, cherts and similar sedimentary rocks in the older Pre-Cambrian would imply an oxygen-rich atmosphere at the time of their deposition, but this may not be the case. Oxidation depends on the chemical form of oxygen, rather than its concentration, and small quantities of chemically active atomic oxygen may well have arisen from photo-dissociation of water, long before life became widespread. Indeed, as we shall see later, the gradual build-up of atmospheric oxygen in late Pre-Cambrian times may well be reflected in the fossil record. Younger Pre-Cambrian rocks, however, contain extensive, oxidized sedimentary iron-ores, and red beds that suggest they formed in an atmosphere containing oxygen, which was presumably produced by photosynthesis of early organisms. The oldest of these red beds are about 1·8 to 2 billion years old, and this may well mark the date of the establishment of an oxygen atmosphere, although this early atmosphere would have contained a much lower concentration of oxygen than does the present atmosphere (Table 1).

THE BIGGEST GAP

We have already seen that the striking contrast between late Pre-Cambrian and Lower Cambrian fossils is not a problem connected with the origin of life. It is in fact a separate and considerable problem, and a number of different suggestions have been made to explain the anomaly. The first type of hypothesis suggests that virtually no Pre-Cambrian organisms existed, the few we know representing exceptional cases. The Lower Cambrian fauna is therefore suggested to have appeared

THE EMERGENCE OF LIFE

as a result of some special, spontaneous creation or a quite unique 'megamutation' or in spores from outer space. There is no evidence which supports such a view, however, and all that we know of the development of life in Cambrian and later times suggests that it is overwhelmingly improbable that Cambrian organisms developed without Pre-Cambrian ancestors. Furthermore, as we have already seen, the most convincing evidences of Pre-Cambrian life have arisen during the last decade from the application of new and sophisticated methods of geochemical investigation, and it would be very rash to suppose that the extension of these studies will not contribute further evidences. The very existence of the few Pre-Cambrian fossils that have been described weighs heavily against the whole viewpoint.

The gap in the chain of life is therefore only an apparent gap, not a real one: a gap in the fossil record, but not in the chain of life. But if so, how can it be explained? One type of explanation suggests that fossils did once exist but they have either subsequently been destroyed by erosion or metamorphism, or they have, as yet, not been discovered. There is clearly some truth in such a view: many fossils may have been destroyed, many others probably are as yet undiscovered. But this certainly cannot be a complete explanation for the rarity of Pre-Cambrian fossils, for there are great thicknesses of unmetamorphosed sedimentary rocks of various types which have been carefully examined in a number of scattered areas. Indeed, in some areas, we have already noted that the palaeontologically defined base of the Cambrian is underlain by great thicknesses of unfossiliferous rocks, apparently unseparated by any physical hiatus. This is true of the western United States, South Australia, and Scandinavia and probably of China, Greenland, and parts of France. It has also been suggested that these presumed organisms may have been confined to fresh water, deep sea, or even terrestrial deposits of which we have no record, but this seems altogether unlikely.

Suppose, however, that the Lower Cambrian fauna developed from ancestors which were soft-bodied, and therefore

THE EVOLUTION OF LIFE

incapable of preservation as fossils. Could this account for their 'sudden' appearance? Many geologists have asked this question, and have suggested various reasons for an earlier absence of hard skeletons. It has been suggested that the Pre-Cambrian oceans lacked available calcium or were too acid to allow the secretion of calcareous skeletons; or that the Pre-Cambrian ancestors were floating or swimming forms for whom the development of hard parts would have been a hindrance. There are strong geological objections to most of these suggested attributes although it is still possible that the basic supposition is correct. Indeed the appearance in the fossil record of relatively complex phyla (arthropods and echinoderms for example) well before such relatively simple forms as sessile corals and bryozoans may lend some support to it, for, if all ultimately arose from a common ancestral stock, corals and bryozoans would seem to have existed for a long period only as soft-bodied forms.

Perhaps the most probable explanation is that at about the beginning of Cambrian times the gradual increase in biologically produced oxygen reached a sufficient level (say 1–3 per cent of its present atmospheric level) to provide an ozone layer which would filter out lethal ultra-violet radiation. This could have allowed the first colonization of surface waters by metazoans and phytoplankton, whose activities would probably have produced a very rapid increase in atmospheric oxygen. A further effect of the increase of atmospheric oxygen would be its availability in concentrations sufficient to support a metazoan level of metabolism.

If this hypothesis should prove to be correct, it does not wholly remove the difficulty of the 'sudden' appearance of hard-shelled organisms in the Lower Cambrian, however, for this is almost as difficult to explain as the sudden appearance of organisms themselves. None of the explanations yet advanced to explain this is completely convincing, yet there are a number of factors to be considered.

As G. G. Simpson has emphasized, it is somewhat misleading to speak of the 'sudden' appearance of most animal phyla in the

THE EMERGENCE OF LIFE

Cambrian. The Cambrian system represents a very long period of time – about 80 or 90 million years – longer than any other system except the Ordovician: longer, in fact, than the whole Cenozoic era. Even the Lower Cambrian probably represents a period of 30 million years. The various phyla do not all appear in the oldest Cambrian strata, but rather 'straggle in' throughout its lower part. The evolutionary rates of change which such appearances would demand from more primitive, soft-bodied forms are not excessive in comparison with those known from later chapters of the history of life.

That so many diverse animal groups should develop hard parts at even a roughly comparable period of time is remarkable, but not unthinkable. The present evidence of Pre-Cambrian fossils at least indicates that a number of lowly forms of life existed before Cambrian times. For how long these existed we do not know, but the earth probably became habitable for them at least as long ago as 3,500 million years, and their record extends almost that far. We know that some forms, such as those from Ediacara, had reached a high degree of complexity and adaptability in such a time, and their diversity and consequent competition would be a selective factor (among others) in their apparently rapid development of hard parts. It is also possible that Cambrian faunas were polyphyletic in their origin, that is, that they arose from more than a single source. Whether or not this is the case, the rapid Cambrian diversification and radiation of the metazoans would almost certainly have been facilitated by the world-wide availability of shallow marine environments, unoccupied by any real competitors, for this diversification was a unique event in the long history of life. The unique atmospheric conditions of late Pre-Cambrian times were probably of great importance in the appearance of Cambrian faunas.

In summary then we may at least claim that the greatest gap of all in the history of life has been narrowed. Numbers of very significant Pre-Cambrian fossils have been found, and, perhaps equally important, we now have a true sense of the time factor involved. The earliest fossils known to us are still complex in comparison with the lowly organisms which must have represented the dawn of life. The ultimate gap will probably never be filled, yet even in those remote and geologically dark ages modern developments in geology, biochemistry, genetics, and physiology throw a gleam of light on the possible origin of life itself. We may never know exactly how this came about, but there seems no reason to postulate that organic change in Pre-Cambrian times was different in kind from the type of change observable elsewhere in the fossil record.

THE ORIGIN OF LIFE

We have already seen that the contribution of fossils to the problem of the origin of life is a limited one. Not only does it raise as many problems as it solves, but by its very nature the fossil record is probably inadequate to record the organisms that represented the first living things. There is, however, another quite different type of contribution that geology can provide, and this concerns predictions about the early conditions of the earth and the seas. It is to this that we must now turn.

Whether, as most geologists and astronomers believe, the earth accumulated by cold accretion, or passed through a molten phase, the environment in which life originated was probably wholly unlike the environment in which it exists today.

Now one of the prerequisites for the origin of life seems to be an environment rich in certain inorganically produced complex organic compounds. In the world as we know it today, we have no evidence of such a synthesis, but in its earlier (not its earliest) condition the situation was probably quite different. While some of the oldest known rocks, ranging from 2,500 to 3,000+ million years old, are represented by water-deposited clastic and chemically precipitated strata, thus indicating weathering and deposition in an atmosphere and hydrosphere, we have also noted that they contain some detrital minerals, such as uraninite and pyrite which show that the atmosphere must have lacked any significant oxygen content. Until recently, it was supposed

THE EMERGENCE OF LIFE

that the original atmosphere was of a reducing nature consisting, not primarily of nitrogen and oxygen, but of water vapour, methane (CH_4), and ammonia (NH_3), plus various other minor constituents, which would include an increasing proportion of nitrogen and carbon dioxide. This view is supported by the spectroscopic demonstration of the atmospheric compositions of other planets (especially the outer ones) which have probably shown relatively much smaller changes in time than has the earth's atmosphere. The atmosphere of Jupiter, for example, consists largely of methane and ammonia. There is now some reason to believe, however, that the primitive atmosphere, though certainly reducing, may not have contained ammonia and methane, but consisted chiefly of water, carbon monoxide, carbon dioxide, nitrogen, sulphur dioxide, and small quantities of other gases. These gases are those most characteristic of volcanoes and thermal springs, and it seems probable that it was from these sources that the earth's original atmosphere was derived, not only by 'degassing' at present levels, but perhaps by some uniquely intense thermal episode early in earth's history. The absence of ammonia is suggested by the predominance of bedded silicates and the rarity of carbonates in the more ancient strata, a situation which would be reversed in the presence of high NH_3 concentration and thus a higher pH of the hydrosphere. Furthermore, a high concentration of methane in the early reducing atmosphere would probably have produced substantial quantities of inorganic carbon by dissociation, but this is not known as a constituent of ancient rocks. Many writers have assumed that atmospheric hydrocarbons would have been formed by the action of water on metallic carbides in the earth. This may or may not be so. Such carbides are not known in the earth's crust today, but they could have existed at earlier dates, and they are, in fact, found in meteorites.

We have already seen that we shall probably never have direct evidence of the origin of life. Fossils are unlikely to help and the transformation of non-living material to living is not

THE EVOLUTION OF LIFE

known to be taking place today. Any hypothesis concerning the origin of life must therefore depend chiefly upon our knowledge of the primitive atmosphere, of existing physicochemical reactions, of genetic and physiological processes, and of the recorded portion of the development of life. All explanations, even though most carefully based on such consideration, will remain only possible explanations, although the construction of experimental models of parts of the process may show that some explanations are stronger possibilities than others. That we shall ever obtain a probable, still less a demonstrably correct, explanation is debatable, but certainly not unthinkable.

Accepting the characteristics of the primitive atmosphere which we have already discussed, what kind of reaction might lead to the development of living organisms? The first step would seem to be the development of inorganically synthesized organic molecules, especially amino acids and carbohydrates. The concentration of such molecules in the seas or lagoons could then ultimately result in the formation of the giant molecules, such as proteins, nucleic acids, enzymes, and genes, which are the building blocks of living things. The absence of both free oxygen and 'life' in this early environment prevented decay. There are certain mechanisms by which these molecular aggregates might be able to duplicate themselves. Both the creation of particular molecular arrangements and this autocatalytic process may well have been assisted by adsorption on mineral particles of clay (especially montmorillonite) and quartz, both of which are very common minerals, and these may also account for the origin of characteristically asymmetrical molecular structure. Such important chemical reactions as photoassociation and polymerization were probably activated by the intense high-energy ultra-violet radiation then reaching the earth's surface, and possibly assisted by lightning discharges. Oparin has suggested a mechanism whereby those organic compounds which include electrically active groups (as proteins do in abundance) would orient water molecules around them. He suggests that if droplets of opposite charges mix, they could precipitate droplets of a complex colloidal aggregate, a

THE EMERGENCE OF LIFE

coacervate, which was capable of absorbing water on its surface and thus might provide a basis for the development of a cellular membrane. This could mark the beginning of individuality.

There is some experimental evidence for the likelihood of at least part of this process. It has been shown, especially by Miller in a classic experiment, that methane, ammonia, hydrogen, and water (all, as we have seen, possibly present in the parent atmosphere), subjected to an electrical discharge (also a frequent atmospheric condition), produce numerous complex organic molecules, including a whole series of amino acids, which are the structural units of protein. Other workers have used other energy sources and other gas combinations, including ones lacking ammonia and methane, and have produced most of the common amino acids which form proteins.

There would be a clear advantage to those molecular aggregates which assisted and supplemented each other's reactions, for their combined activity could then be more effective than when it occurred singly. By something analogous to natural selection, these could ultimately develop into organized aggregates – 'living things' by any standard.

Such organisms would probably have been heterotrophic, that is 'other feeders', which, in the absence of oxygen, lived by anaerobic fermentation, involving the breakdown and build-up of hydrocarbons from the reservoir of organic compounds from which they arose. But this reservoir was clearly limited, and, as Simpson and others have pointed out, its ultimate exhaustion destroyed for ever the conditions required for the spontaneous origin of living matter from non-living. It must therefore be supposed that some 'mutational' change gave rise to the capacity of some organisms to synthesize their own food products (autotrophs) from more widely available inorganic sources, and thus to the development of photosynthesis. Photosynthesis, a process characteristic of all plants, involves the production of various organic compounds from water and carbon dioxide, in the presence of energy provided by sunlight. It is accompanied by the release of oxygen. Such a process would be dependent upon the existence of carbon dioxide, now available from the

THE EVOLUTION OF LIFE

fermentation of early organisms, and would itself lead to the slow development of an oxygen-bearing atmosphere. This oxygen atmosphere not only provided the possibility of the later development of aerobic respiration but also led to the formation of the ozone layer of the atmosphere, thus shielding the earth from the previously intense ultra-violet radiation, while transmitting visible light which provided the chief source of energy for plants, and ultimately for animals. The later development of respiration was a major event in the history of living things, not only because it provided further CO_2 for the photosynthesis cycle, but also because it produces over 30 times more energy than does fermentation – energy which was thus available to new groups of organisms for growth and movement.

Preston Cloud has pointed out that in the probable early absence of oxygen-mediating enzymes, the oxygen-releasing autotrophs would have required some means of disposing of the potentially poisonous oxygen they produced. Cloud suggests that the banded hematitic iron strata, uniquely characteristic of the period of earth history between about 2 and 3 billion years ago, acted as a physical oxygen acceptor and control. He suggests that ferrous iron in solution may have been oxidized to ferric oxide, and then incorporated in these rocks by precipitation. This regime, if it existed, would have ended with the emergence of oxygen and peroxide-mediating enzymes, which then became available to early photosynthesizing autotrophs.

The slow build-up of oxygen would also have allowed the emergence of advanced types of cells, with nuclear walls and well defined chromosomes, which were capable of cell division and ultimately sexual reproduction.

The whole foregoing hypothesis (or any other of a similar type) involves such a long succession of ifs, buts, and perhapses that many people have rejected any such natural explanation of the origin of life as so statistically improbable as to be impossible. But improbability is by no means synonymous with impossibility. The countless billions upon billions of atoms available on the earth over a period of perhaps 2,000 million years greatly increase the probability, simply because of the

THE EMERGENCE OF LIFE

increase in the possibility of its occurrence. Furthermore, we must not forget that the event took place in a primitive environment totally different from that in which we live today.

By such a succession of events, then, it is possible that living things developed. The hypotheses that have been put forward are no more than intelligent guesses, possible certainly – but by no means proved. It is in fact difficult even to visualize the full range of reactions and events which could have led to the wonderful and complex process of life in which we, for a time, have a part. At present we can only speculate. Indeed we may never know the true process by which life arose – but life did originate and at a much later date a record of its development was preserved. It is to that record that we must now turn.

Chapter 5

THE HEY-DAY OF MARINE INVERTEBRATES

CREATURES OF THE CAMBRIAN SEAS

WE have already seen something of the faunas of the Lower Cambrian, and noted that they consisted very largely of marine invertebrates. Before we look at these in more detail, however, we must remind ourselves of two things. Firstly, the geography of the past was very different from the geography of today. There was nothing fixed, nothing constant in the position of land and sea, mountain and plain. As a matter of fact, it is only because this was so that we have any record of the most ancient life at all (for we are now able to collect marine fossils in land areas). Secondly, not only the broad framework but also the whole appearance and character of the earth was completely different. So far as we know, not a single animal moved across the land, no animal flew through the empty air, and probably none dwelt in the streams and lakes. Even the earth itself was utterly barren – a wilderness of bleak, naked rock and boulders, and dust. There were no trees, no shrubs, no grasses, no flowers. It was only, perhaps, near the margins of the seas that we might have seen small patches of green algae, encrusting the rocks. No words can quite convey the world-wide emptiness and desolation of those distant days.

But in the seas things were very different, for it was here that earlier actors played their roles in the great drama of life. It is within their shallow waters, some 600 million years ago, that we get our first full glimpse of the drama – a drama, in one sense, by now well into its final act, for its earliest scenes (now lost to us) probably took place two or three thousand million years before this.

Sponges, jellyfish, gastropods, brachiopods, worms, arthropods, echinoderms – these are the creatures which thronged the

THE HEY-DAY OF MARINE INVERTEBRATES

shallow seas at the dawn of the Cambrian. There were almost certainly other things living with them whose remains have not yet been discovered in the Lower Cambrian, but which are known either from older rocks (such as primitive plants) or from somewhat younger ones (protozoans, for example). It was these groups which gave rise to the host of marine invertebrates that dominated the life of the Palaeozoic Era.

The sponges

The sponges (*Phylum Porifera*) are the simplest of all the multicellular animals. One of the most striking indications of this simplicity is their amazing power of regeneration. Sponges may be crushed and the cells sieved through a fine silk mesh under water, whereupon the cells will gradually aggregate and form masses from which new sponges ultimately develop. This raises one of the most perplexing questions in zoology – the question of what constitutes an individual.

There are some who believe that in the sponges which contain more than one excurrent opening it is these openings themselves that constitute the centres of the individuals, so that the sponge is really a colony. From this viewpoint boundaries between individuals are indefinite, since all the excurrent openings are part of one system of canals. Such 'indefinite individuality' is not a characteristic of the higher animal groups, because it is the result of a low-level organization unique to the sponges – the development of cell differentiation without any appreciable cellular coordination (although the flagellar beat of sponges has been so interpreted by some).

Most living sponges, and all those of the Cambrian, are marine, and they live in a fixed position. They vary widely in colour, shape, and size, the largest being over six feet in diameter. It is only just over a hundred years ago that they were finally shown to be animals, rather than plants. They contain collared flagellate cells, which strongly resemble some colonial flagellate protozoans (see p. 125), and also employ intracellular digestion. These similarities are regarded by some as an indication that sponges arose from these protozoans.

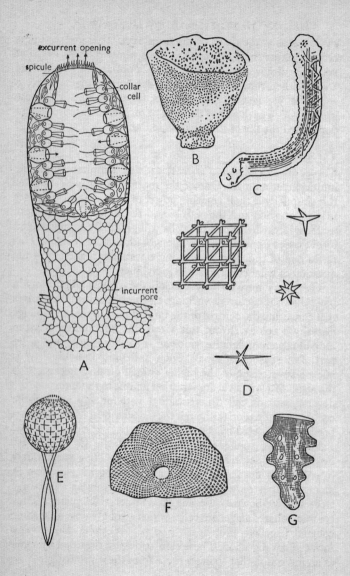

100

THE HEY-DAY OF MARINE INVERTEBRATES

The body of the sponge is essentially an animated sieve. The body wall is perforated by large numbers of incurrent pores and canals, through which water flows into a central cavity and thence flows out through one or more openings (*osculum /a*) at the top of the sponge. Movements of the whip-like flagellate cells produce a continual flow of water, which provides oxygen and food (assimilated by intracellular digestion) and removes waste products. Reproduction may be asexual or sexual, and sponges display a free-swimming larval stage. The adults, however, possess no movable parts, appendages, or organs. The body may be supported by a skeleton of organic fibres or of minute intricate siliceous or calcareous spicules of varying shapes, or of both. Sponges are known from some of the oldest fossiliferous rocks and, although they are widespread and not uncommon throughout geological time, they are a conservative and relatively unimportant group. It seems likely that they represent an evolutionary side-line, which did not give rise to the other metazoans.

The older fossil sponges were a mixed group, some not greatly different from living forms, other now long extinct. Many were supported by a delicate net-like skeleton of silica, similar to that of the beautiful living glass sponge (*Euplectella*, or Venus's flower basket).

In some forms the spicules were discrete, in others they were fused together to form a more or less rigid framework or 'skeleton'. Isolated spicules are quite common in certain rocks, and are amongst the most delicate and beautiful of all microfossils. Some of the many variations in form are shown in Fig. 7 D.

FIG. 7 Sponges – living and fossil.
A A simple sponge, showing general structure (*after Buchsbaum*); B–C Living sponges: B a horny sponge from the Mediterranean, C a deep-sea glass sponge, *Euplectella*, 'Venus's flower basket'; D Sponge spicules, showing variation in shape (highly magnified); E–G Palaeozoic sponges: E *Protospongia*, from the Cambrian, height 4 inches, F *Receptaculites*, a sponge-like fossil from the Ordovician, diameter 3–4 inches, G *Hydnoceras*, from the Devonian, height 8 inches.

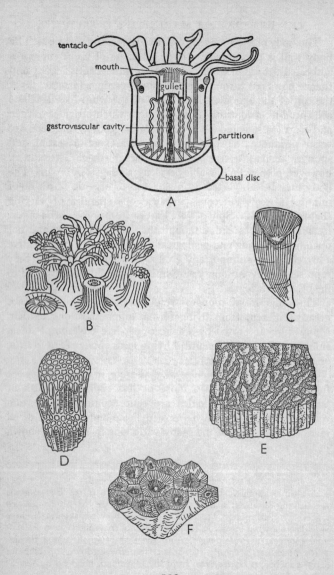

THE HEY-DAY OF MARINE INVERTEBRATES

Such sponges were not, however, the most distinctive of Cambrian times, for one group, the pleosponges or archaeocyathids (Fig. 6 H), are confined to the Cambrian, and are so common that they formed extensive reefs in many parts of the world, from North America through the Mediterranean to Asia, Australia, and Antarctica. Almost 500 species assigned to over 90 genera have been described. They often reached almost a foot in length, and they had a vase-like calcareous skeleton consisting of a double perforated wall with intervening radial partitions. Although they are generally regarded as sponges, the interesting thing is that they also show quite considerable resemblances to the calcareous corals: a resemblance which has led some authors to regard them as ancestral to both corals and sponges.

The coelenterates

Above the sponges, which were fixed dwellers on the floors of the Cambrian seas, there floated the delicate polyps of jellyfish. These beautiful creatures are amongst the softest of all soft-bodied animals, and although they are therefore uncommon as fossils, they are known even from these ancient Cambrian faunas. They belong to the *Phylum Coelenterata*, a group which also includes the sea anemones, corals, sea fans, sea pens, and the fresh water *Hydra*. The coelenterates are aquatic animals, most of them marine, either solitary or colonial, and they have a sac-like body cavity, with a ring of tentacles surrounding the single opening or mouth (Fig. 8A). In spite of the delicate flower-like appearance of some of their members, coelenterates are formidable creatures, capturing and eating not only microscopic organisms but also molluscs, crustaceans, and fish, which they paralyze by means of stinging cells.

FIG. 8 Coelenterates.
A Structure of a sea anemone (*after Buchsbaum*); B Living corals; C–F Palaeozoic corals (all about half natural size); C *Streptelasma*, Ordovician, D *Favosites*, Ordovician–Permian, the honeycomb coral, E *Halysites*, Ordovician–Silurian, the chain coral, F *Lonsdaleia*, Carboniferous.

The coelenterate body may be of two kinds. The *polyp* (Fig. 8 B) is a sac-like body, closed and attached to some object at one end, and having a mouth and tentacles at the other (e.g. sea anemones and corals). The *medusa* is a free swimming, umbrella-shaped form, with a ring of delicate tentacles from the lower (concave) surface (e.g. jellyfish). Both forms represent a fundamentally similar structural plan and alternate in the life cycle of some species. In some coelenterates, such as corals, the body is supported by a skeleton. The mouth serves both for the intake of food into the digestive cavity (coelenteron) and also for the expulsion of waste and sexual products. Coelenterates have a primitive nervous system and some have eyespots, but they lack circulatory, respiratory, and excretory organs.

The jellyfish are the first known representatives of the coelenterates, and although they are rarely preserved as fossils, they, like the sponges, have persisted to the present, with little, if any, modification in structure, as important though inconspicuous members of the world's seas. The clue to the conservative history of both groups lies in their successful adaptation to widespread, though distinct, environments. Sponges are as much a model of economy and simplicity of structure in their adaptation to life on the sea floor as are jellyfish to life in the open seas. Neither environment has changed very much, both have remained widespread; and with their persistence the sponges and jellyfish have also persisted. It is one of the few valid generalizations that can be made of the history of life that it is the generalized types of organisms that tend to persist and the more specialized that perish.

The molluscs

In the rock pools along the shores of Cambrian lands and in the waters of the seas, the slow-moving snails browsed, much as their whelk and winkle descendants do today. These gastropods are the oldest representatives of one of the most varied and successful of all animal groups (the *Phylum Mollusca*), which also includes such apparently dissimilar animals as slugs, oysters,

THE HEY-DAY OF MARINE INVERTEBRATES

scallops, clams, chitons, cuttlefish, the octopus, and the nautilus. Some molluscs are indirectly familiar as a source of food (to those who can eat them) and countless forms abound on the seashores the world over. They vary greatly in size, the smaller ones being microscopic, while the giant squid may reach a length of well over seventy feet, and the clam *Tridacna* may weigh over 500 pounds.

In spite of their external differences, all molluscs share a fundamentally similar body plan (Fig. 9 A–D). They have a soft, unsegmented body, typically with an anterior head, and a large viscreal mass supported on a fleshy muscular foot. The body is largely surrounded by a thin fleshy layer (the mantle), which secretes a shell of various types in most members of the phylum. In clams and oysters the shell consists of two valves, while in snails it usually consists of a single coiled shell. The chitons are protected by a series of eight movable plates, while in cuttlefish and squids the shell is completely internal. A few naked forms bear no shell, either internal or external. The sexes are usually distinct. The fundamental body plan, which distinguishes the molluscs from all other groups of animals, is best developed in the chitons. In more specialized members of the phylum, such as clams and squids, some of the typical molluscan features may be strongly modified, or even lost altogether.

Of the five classes of molluscs only three are commonly found as fossils. The two remaining groups (the *Amphineura* or chitons and the *Scaphopoda* or tusk shells – see Fig. 9) are known as fossils, but are rare.

The gastropods, which include both the familiar snails and the slugs, are the most widely adapted and diverse as well as the oldest group of molluscs. Although their earliest members were exclusively marine, the 20,000 modern species inhabit a very wide variety of marine, fresh-water, and terrestrial environments, from the tropics to the subpolar and desert regions and from 18,000 feet above sea level down to 17,000 feet below it. The body has a distinct head (often bearing stalked eyes and other sensory organs) and a well-developed foot, and is usually

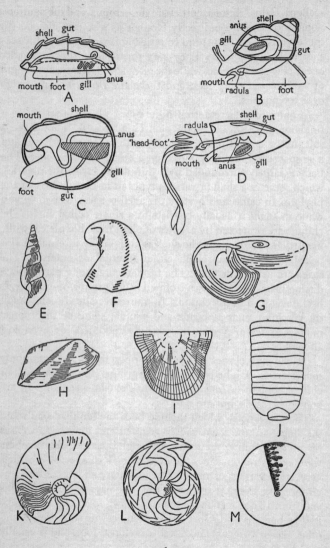

106

protected by a coiled unchambered, calcareous shell of variable form and size. In some species the shell aperture is closed by a small plate, the operculum. The gastropod mouth is equipped with a formidable chitinous band (the radula) which is armed with rows of minute teeth. These teeth, which may number as many as 750,000, form a very efficient rasping structure, which, in such genera as *Natica*, is strong enough to drill holes in other shells.

Some gastropods are predatory carnivores, but most are herbivores, grazing on the lowly plants of the oceans. They are amongst the most abundant fossils of the Cambrian, where they are represented by small, more or less conical forms (Fig. 6c). Complex coiling and the growth of heavy calcareous shells were later developments.

The worms

Marine worms (of the *Phylum Annelida*) seem to have been as widespread in Cambrian seas as they are today. The Middle Cambrian Burgess Shale fauna (see Pl. 3 (*b*)), for example, contains almost perfectly preserved specimens, but in general their soft bodies make worms unlikely objects for preservation as fossils, and they are more usually represented by fossil tracks, burrows, tubes, and scolecodonts (the minute chitinous jaw components of polychaete worms).

The annelids include a variety of different forms, such as the common earthworm, and leeches, as well as the marine polychaetes (*Nereis* and other genera) and the tube-dwelling

---

FIG. 9 Molluscs.

A–D Structural similarities between: A chiton, B gastropod, C pelecypod, D cephalopod; E–G Palaeozoic gastropods (all half natural size): E *Loxonema*, Ordovician–Carboniferous, F *Platyceras*, Silurian–Permian, G *Maclurites*, Ordovician, basal surface uppermost; H–I Palaeozoic pelecypods (about natural size): H *Goniophora*, Silurian–Devonian, I *Dunbarella*, Carboniferous; J–M Palaeozoic cephalopods showing increased complexity of suture line; J *Endoceras*, Ordovician ($\times \frac{1}{3}$) (showing siphuncle), K *Agoniatites*, Devonian ($\times \frac{1}{4}$), L *Goniatites*, Carboniferous ($\times 2$), M *Medlicottia*, Permian ($\times 1$).

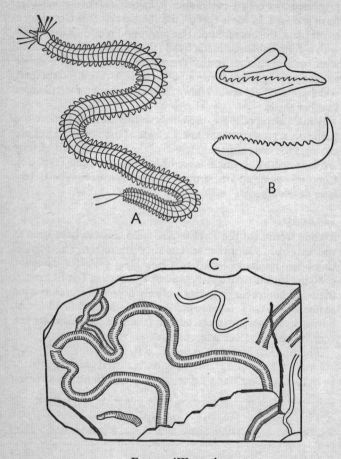

FIG. 10 'Worms.'

A *Nereis*, a modern polychaete worm; B Scolecodonts, the minute chitinous jaw components of fossil worms (greatly magnified); C fossil worm-borings in a rock (natural size).

worms. Many are beautifully coloured and delicately 'plumed'. The worm-like body has a mouth and anus at opposite ends, and in most forms is divided into a number of transverse segments (or 'rings' – hence the name 'Annelida'), which frequently bear small muscular processes used in both locomotion and respiration. Rows of bristles may also be present along the tubular body. Annelids have well-developed nervous and excretory systems and, in spite of their apparent simplicity, they are relatively complex organisms.

Although they are not conspicuous fossils, worms have probably played in the past the same important part that they now play in the development both of soils and of unconsolidated sediments on the sea floor. This sediment is passed through the alimentary tract of worms, which extract food particles from it, the sediment itself being modified both by grinding and by the chemical reactions of the worms' digestive processes. Charles Darwin showed that some land areas have a worm population that must annually transport eighteen tons of soil per acre to the surface, while similar calculations show that the marine worms of the Northumberland coast carry 3,147 tons of sediment per acre to the surface each year. The entire upper two feet of the coastal sediment would thus pass through the worms in a period of two years. Such modification of sediment is of the greatest importance in the geological processes of rock formation.

The echinoderms

The oldest representatives of the echinoderms are also found in the Cambrian. The *Phylum Echinodermata* includes the living starfish, brittle stars, sea urchins, sea lilies, and sea cucumbers, as well as the extinct blastoids, cystoids, and edrioasteroids. All echinoderms are marine but some lead a fixed sedentary life while others are free moving and swimming forms. These apparently diverse animals share a number of important structural features. They all tend to be more or less spiny, and the body is encased either by a calcareous test of globular or discoidal shape (as in sea urchins), or by a leathery

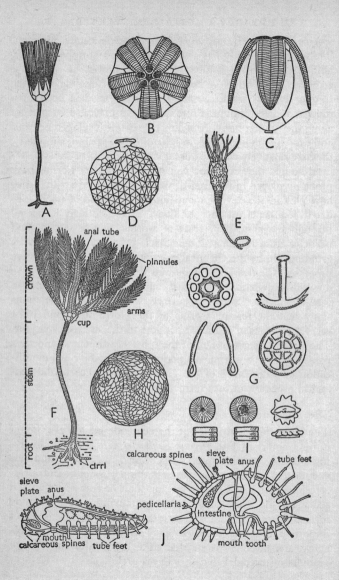

THE HEY-DAY OF MARINE INVERTEBRATES

skin having a star or ellipsoidal shape, studded with small calcareous plates (as in starfish). They usually exhibit a fivefold radial symmetry and the sexes are normally separate. They have a simple digestive tract, a localized nervous system, a primitive circulatory system, and a well-developed water vascular system, which consists of a system of canals containing fluid, often terminating in hollow extensible tube feet which are used in locomotion and food gathering. A number of echinoderm body-features are similar to those found in chordates, and there is such a striking resemblance between some larval forms and the larvae of hemichordates that some students have concluded that the echinoderms may be close to the forms from which the chordates originated (p. 162).

Echinoderms include some of the most exquisitely coloured and delicate of all animals, yet for all their apparent tranquil beauty, they are amongst the most aggressive scavangers and predators in modern seas. They are very widely distributed in both depth and latitude, and many live in colonies of enormous numbers of individuals.

The Cambrian echinoderms include representatives of archaic and now extinct groups. The edrioasteroids (Fig. 11H) were globular and rather irregular forms, with five sinuous 'arms' (bearing food grooves) radiating from the mouth. They persisted until Carboniferous times, but they were never important members of the faunas of the Palaeozoic. With them

---

FIG. 11 Echinoderms.
A–C Blastoids, *Pentremites*, Carboniferous: A restoration of animal, showing stem supporting bud-like calyx, which bears arm-like pinnules (about half natural size), B–C upper and lateral views of calyx, showing petaloid ambulacral areas ($\times 4$) (*after Bather*); D *Echinosphaerites*, an Ordovician cystoid (natural size); E *Macrocystella*, a Cambrian eocrinoid (natural size); F *Botryocrinus*, a Silurian crinoid, showing general structure (natural size) (*after Bather*); G Holothuroidean spines (greatly magnified); H *Edrioaster*, an Ordovician edtrioasteroid (natural size); I Crinoid columnals from stems (natural size); J Starfish and sea urchin, cross sections to show similarity of general structure (*after Buchsbaum*).

THE EVOLUTION OF LIFE

in Cambrian times there existed the eocrinoids (Fig. 11 E), a small group of creatures confined to the Cambrian and Ordovician. They consisted of a calyx with branching arms, which was attached to the sea floor by a short stalk. These ancient echinoderms contain features which are characteristic of both the later crinoids and cystoids (see p. 134), and it seems probable that they are the ancestors of these groups.

The sea cucumbers (holothuroids) are elongated sac-like animals, with an anterior mouth surrounded by a ring of tentacles. Their leathery skin contains curiously varied minute plates (Fig. 11 G) which are not uncommon as fossils. The animals lead a sluggish or burrowing life on the sea bottom, and they too are known from Cambrian strata.

The brachiopods

The most characteristic of all the animals of the Cambrian seas were not the sponges, jellyfish, gastropods, echinoderms, or worms. They were the lowly inarticulate brachiopods and the ruling trilobites. Brachiopods (*Phylum Brachiopoda*) are small, bottom-dwelling marine 'shellfish'. Their shells consist of two valves, which are bilaterally symmetrical, although they usually differ in shape and size. These valves, which completely surround the soft parts of the animal, may be hinged with teeth and sockets (these are the so called articulate brachiopods) or they may be held together only by muscles (as in the inarticulate brachiopods). The animal is attached to the sea bottom by a short, fleshy stalk or pedicle, which usually perforates the larger valve (Fig. 12). The other soft parts consist of two long coiled structures (the lophopore), which bear ciliated tentacles that circulate a flow of water into the body for respiration. From this water food particles are flushed into the mouth. A small 'heart' and blood vessels are present, together with excretory organs and a simple nervous system. The sexes are distinct.

The inarticulate brachiopods, of which *Lingula* is an example, are the more primitive, and it was these which existed in great numbers in the Cambrian seas. Their shells are composed of calcium phosphate and/or chitin, and they have a dark shiny surface (Fig. 6A).

The articulate brachiopods represent the more advanced and larger division of the brachiopods, the earliest members of which appeared towards the close of Cambrian times. Their hinged shells are composed of calcium carbonate and are generally light grey in colour, exhibiting much greater diversity than the inarticulate forms.

Although less conspicious than the trilobites, the brachiopods rival them in diversity and abundance in the Lower Palaeozoic, and are perhaps the most characteristic of all the Palaeozoic invertebrates. Living brachiopods are a small group of about 200 species, but they were almost universal in earlier seas. Their oldest representatives, as we have already seen, were the inarticulates of the Lower Cambrian, a group of small, dark unhinged forms, with chitinous and phosphatic shells. Within this group there was appreciable variation in form – some were oval or pear-shaped or circular and flattened, others were round or conical in shape and there were many minor differences in structure, especially in the arrangement whereby the anchoring pedicle perforated the shell. Some members of this lowly group of inarticulate brachiopods have persisted largely unchanged to the present and their development is therefore conservative and their history unspectacular. This does not imply that they have been a 'failure'; on the contrary, their very lack of radical change is an indication of their success in adaptation to the shallow environment in which they live. One living genus, *Lingula*, has an unbroken history going back 500 million years into the Ordovician. Now how is it that such a group, let alone a genus, can survive so long unchanged? Certainly the answer is not sheer weight of numbers or abundance, for the group has been relatively inconspicuous in post-Cambrian times. Nor is the answer to be found in an elaborate degree of body structure which is capable of withstanding both the wiles and competition of predators and competitors and the vicissitudes of changing environments. As in the sponges and jellyfish, the real clue to the survival of *Lingula*

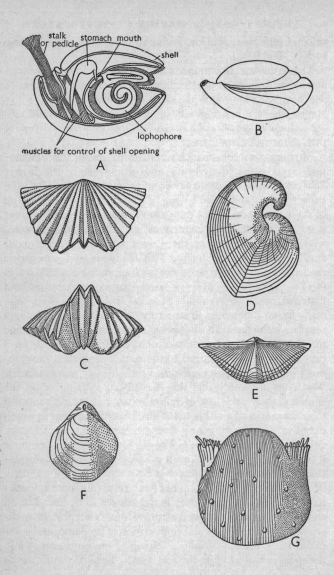

THE HEY-DAY OF MARINE INVERTEBRATES

and its allies seems to lie, not in their complexity, but in their simplicity. When we find fossil inarticulate brachiopods in post-Cambrian strata they are almost always found in dark shales and this (as well as their present habits) suggests that they have remained unspecialized because they are well adapted to an environment (muddy, warm coastal waters) that is very widespread. The living *Lingula* exhibits a quite remarkable degree of tolerance to salinity changes, to changing composition of water, and even to quite considerable periods of exposure on tidal flats.

The trilobites

The trilobites were the masters of the Cambrian seas. Complex in structure, enormously varied in form, swarming in countless numbers, they dominated the fauna of the ancient seas, and by late Cambrian times there were no fewer than 700 genera. The trilobites are members of the arthropods, the most varied and by far the largest of all the phyla. The *Phylum Arthropoda* contains about three-quarters of all known animal species. As if these qualifications were not in themselves sufficient, many arthropod species are enormously abundant as individuals. The arthropods include such familiar representatives as the shrimps, crabs, lobsters, barnacles, spiders, scorpions, ticks, centipedes, millipedes, and insects, as well as the extinct trilobites and eurypterids.

The arthropod body and limbs are modified to different modes of life in a great variety of different environments, ranging from the depths of the ocean through most terrestrial and fresh water habitats into the air. Fundamentally the body is composed of a number of segments with jointed appendages,

---

FIG. 12 Brachiopods.
A Structure of an articulate brachiopod; B brachiopod shell in same relative position as A; C *Platystrophia*, Ordovician–Silurian; D *Conchidium*, Silurian; E *Mucrospirifer*, Devonian; F *Composita*, Carboniferous–Permian; G *Linoproductus* Carboniferous–Permian. (All approximately natural size)

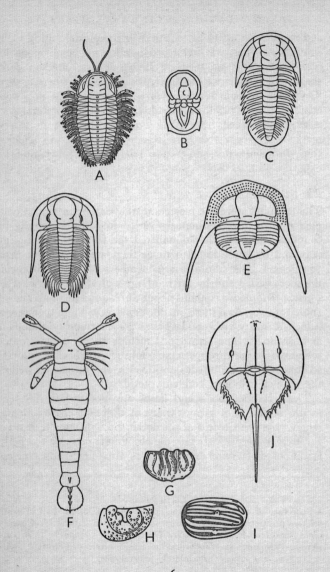

THE HEY-DAY OF MARINE INVERTEBRATES

and is covered by an external skeleton of varying rigidity. This is periodically shed, during moulting. The digestive, circulatory, nervous, and reproductive systems are all well-developed.

Of the five most familiar classes of arthropods, only two (the crustaceans and the insects) are common as fossils, although representatives of the remaining three (including spiders, scorpions, eurypterids, centipedes, and millipedes) are also found.

The crustaceans include the modern crabs and lobsters and are the closest class to the extinct trilobites. The trilobites are elegant and well-loved fossils, having bodies of a variable number of segments, covered by an external skeleton of chitin. As their name implies, they were, like Caesar's Gaul, divided into three parts (see Fig. 13A). They exhibited a considerable size range (from about $\frac{1}{8}$ inch long to over two feet) but most of them were two or three inches long.

During the Lower Palaeozoic the trilobites were supreme and in the oldest fossiliferous rocks trilobites are usually the most common fossils, larger and more highly developed than any of their rivals. Their remains are rather more common than we might expect, because the creatures cast off their encasing skeletons at successive periods of growth. Most Cambrian forms were relatively small (two to four inches was average) although others were minute and some much larger. *Paradoxides*, a common Middle Cambrian fossil in parts of Wales and around Boston, Massachusetts, had a length of about eighteen inches (Fig. 13D).

FIG. 13 Arthropods.

A–E Trilobites: A a restoration of *Triarthrus becki*, a trilobite from the Ordovician of New York (*after Beecher and Raymond*); B–D Cambrian trilobites: B *Agnostus*, Cambrian (×4), C *Bathyuriscus*, Middle Cambrian, D *Paradoxides* Middle Cambrian, length up to 18 inches, E *Cryptolithus*, Ordovician (×2) (all natural size, except as stated); F *Pterygotus*, Ordovician–Devonian, a eurypterid, up to 7 ft. long; G–I Palaeozoic ostracods (greatly magnified); G *Tetradella*, Ordovician–Devonian, H *Hollina*, Devonian, I *Glyptopleura* Carboniferous–Permian; J The living king crab, *Limulus*, a 'living fossil', length 18 inches.

THE EVOLUTION OF LIFE

Cambrian trilobites, though enormously varied, display a number of characteristics which distinguish them from most later forms, and we can regard them as made up chiefly of three broad types. The smallest are the minute agnostids and eodiscids (Fig. 13 B) – rarely more than ¼ inch in length, having broadly similar and more or less smooth heads and tail shields and only two or three thoracic segments. Most, but not all, of them were blind and the group as a whole was widespread during Cambrian times. The second conspicuous group (quite heterogeneous in any taxonomic sense) was a much larger one and includes the host of rather long, shaggy trilobites, having large crescentic eyes, prominent genal and pleural spines (Fig. 6 D, E), many thoracic segments, and a small pygidium. The remaining group includes forms which tend to have a rather oval and smooth outline (though sometimes with genal spines), many of which were blind, distinctly trilobed with well-marked glabella lobes, but almost totally lacking in surface ornamentation and spines (Fig. 13 C).

Such complexity and variety as this is a forceful reminder of the widespread adaptation of the group, even in Lower Cambrian times, and clearly suggests a long Pre-Cambrian history.

Such were the trilobites – masters of the Cambrian seas. But they were not the only arthropods in those early days. There were other crustaceans in Middle Cambrian times about which we know a good deal, as a result of a remarkable fossil discovery by the great American palaeontologist, Charles D. Walcott, in the Canadian Rockies above the town of Field, British Columbia. Walcott discovered above Burgess Pass on the steep flanks of Mt Wapta, a band of dark shale (the Burgess Shale), about seven feet in thickness, which has yielded a fauna including more than seventy-five genera. The most remarkable feature of this fauna is its exquisite state of preservation, for the impressions of even the most delicate soft parts of the organisms are faithfully preserved. Many of the organisms, in fact, are entirely soft-bodied, and are quite unknown apart from this isolated locality. The state of preservation and the lithology of the shale suggest that it accumulated in a stagnant unoxygenated embayment or depression, from which scavengers were absent. Jellyfish, worms, sponges, sea cucumbers, trilobites, and other small bivalved and shrimp-like crustaceans are included in the fauna, and in some of them impressions of even the internal organs are clearly preserved (Plate 3 (*b*)).

The distribution of Cambrian faunas

One other general feature of Cambrian marine invertebrates that remains to be mentioned concerns faunal provinces. Almost all living marine invertebrates, like other animals and plants, have a more or less restricted geographical distribution, and a similar type of restriction may be recognized from Cambrian times onwards. The similarity did not, of course, extend to the pattern of these faunal provinces, for the extent of seas of Cambrian times was quite different from those of today. This distribution is particularly well shown in Cambrian trilobites, certain genera of which are frequently confined to particular provinces.

In Middle Cambrian times there are very marked differences between, for example, the trilobite faunas of the eastern and western United States. The western (Pacific) Cambrian realm is characterized by large-tailed trilobites (such as *Bathyuriscus*) which are absent from the Atlantic province, whose typical *Paradoxides* trilobites are likewise not found in the West. This *Paradoxides* fauna may be traced through New England, Newfoundland, and Nova Scotia into Europe, where the Scandinavian and Welsh faunas, for example, are very similar. It is very striking that both in Scotland and in the southern Appalachian region of the Atlantic province, however, there is no trace of *Paradoxides*, for in those areas Pacific-type trilobites were living in Middle Cambrian times. It is therefore supposed that there was free migration along shallow seas which connected these areas. The presence of different faunas within a limited geographical area probably indicates that intermingling of the two faunas was prevented by the deeper waters of the Cambrian oceans.

THE EVOLUTION OF LIFE

It may be, however, that another factor was also important in this distribution. A recent detailed study showed that Upper Cambrian olenid trilobites, although they had a world-wide occurrence, are apparently confined to dark-shale-limestone-depositing areas, bordering ancient geosynclines (elongated downwarped belts of more or less continuous subsidence and sedimentation) now represented by Palaeozoic mountain belts. Quite different faunas occupied the shallow water around each continental area. These faunas are more varied than those of the olenid provinces (180 genera as against 80) where the relative uniformity of physical conditions seems to have led to a more restricted evolutionary development.

Such were the faunas of the Cambrian seas – a mixture of archaic forms dominated by the trilobites and inarticulate brachiopods, but including primitive echinoderms, gastropods, worms, and coelenterates. The general character of this fauna clearly implies that marine plant food must also have been abundant, especially the algae (seaweed and the minute diatoms, p. 147), the grass of the oceans. Such soft-bodied plants are rare as fossils, but the twenty species of algae described from the Middle Cambrian Burgess Shale are an indication that they were abundant in ancient seas. Some algae were lime-secreting, and their remains are also known from the Cambrian. The tiny diatoms, which are the ultimate basis of the great food chain of the present oceans, are not known as fossils until Mesozoic times, but it seems at least probable that they existed as soft-bodied forms in much earlier periods.

LIFE OF THE MIDDLE PALAEOZOIC SEAS

This balanced assemblage of ancient organisms persisted for a period of 90 million years until Late Cambrian and Ordovician times, when it underwent a radical transformation. This transformation was twofold in character, for it involved both the decline of some of the older groups (such as the trilobites and inarticulate brachiopdos), and the appearance and expansion of a number of completely new forms (protozoans,

THE HEY-DAY OF MARINE INVERTEBRATES

corals, articulate brachiopods, bryozoans, pelecypods, cephalopods, eurypterids, ostracods, crinoids, blastoids, echinoids, starfish, graptolites, and fish), some of which dominated the faunas of what we might loosely call the Middle Palaeozoic (the Ordovician, Silurian, and Devonian) (Fig. 5).

The geography of this period of about 175 million years was startlingly different from that of our own times. The climates, the shapes and positions of land and sea, perhaps even the position of the poles, bore little resemblance to those of today. Lower Palaeozoic times were characterized by widespread warm or equable shallow seas. Limestones with reef-building corals are found in polar regions and continuous down-warping of long submarine troughs in certain areas (the 'geosynclines' of the Appalachians, Scotland, Wales, and Norway produced thicknesses of up to seven miles of sedimentary rocks). Active volcanoes studded these regions and there were continual and extensive advances and retreats of the ancient seas. We know little about the physiography of the land during these times, although locally there seem to have been deserts. Mountain building during late Silurian and Devonian times accompanied the spread of lands with warm, semi-arid climates, and heavy seasonal rains. Great thicknesses of red sediments (the 'Old Red Sandstone') accumulated in inland mountain basins and coastal deltas in north-western Europe and North America, while corals flourished in the warm seas. By late Devonian times there were extensive forests in some areas. It was against this background, as we shall later see, that the fishes rose to a position of dominance and the amphibia clambered ashore.

The trilobites

We have already seen that some of the ancient Cambrian creatures, such as jellyfish, sponges, and sea cucumbers, persisted through subsequent periods with little change. The change in other groups, however, was drastic. Consider, for example, the trilobites. Their history is partly, but only partly, told in their numbers. By Late Cambrian times at least 700 genera were in existence, while at the close of the Ordovician there were 200 genera and 1,200 species, most of them quite different from those of the Cambrian; but from then on they were represented by a steadily diminishing number of genera, except for a slight resurgence in the Devonian, after which they dwindled on until Permian times. But in spite of this steady decline, the Ordovician history of the trilobites is scarcely less spectacular than that of the Cambrian, although the general appearance of the two faunas is quite different. Most of the typical Cambrian forms were now extinct, and one of the most distinctive Ordovician groups included the trinucleids with a broad, pitted, flattened margin around the head shield (Fig. 13E). Ordovician trilobites display a good deal more variation in general form than do those of the Cambrian, and they underwent a host of different evolutionary trends. Many developed bizarre spines and nodes, others tended to develop enormous globular eyes with over 1,000 lens-like facets, still others were blind, some developed shovel-like snouts (presumably an adaptation for life in muddy sediments), others lost all trace of segmentation in the cephalon. Most continued to be rather small, but one giant reached a length of well over two feet. Amongst these more extreme developments, other forms remained less modified and this more conservative stock not only persisted with the more highly modified forms into the succeeding Silurian and Devonian, but also survived them by many millions of years to the end of the Palaeozoic. This, as we have already seen, was no accident, for time and time again in the history of life it is the unspectacular, unspecialized forms which have survived and the highly modified and adapted which have not.

Something of the enormous range of variation which the trilobites displayed in their day is shown in Fig. 14, and it is always tempting to interpret the significance of the more striking modifications in relationship to the way of life of the trilobites. Such interpretations raise a number of problems. Many authors have pictured the history of animal groups in terms of the life history of individuals, and have attempted to read into the development of spines and ornament signs of

THE HEY-DAY OF MARINE INVERTEBRATES

racial degeneration and senescence. This, however, is almost certainly much too naïve an interpretation. We shall discuss the general problems of survival and extinction in a later chapter (Chap. 8), but we are already in a position to observe

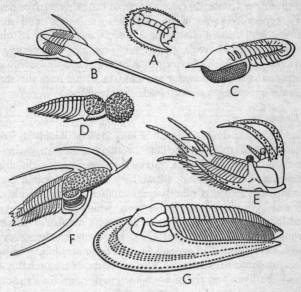

FIG. 14 Trilobites, showing variation in form and probable modes of life.

A Larva of *Acantholoma*, Silurian (×20); B *Lonchodomas*, Ordovician (natural size); C *Symphysops*, Ordovician (×1½); D *Staurocephalus*, Silurian (×1½); E *Ceratarges*, Devonian (×2); F *Teratorhyncus*, Ordovician (×2½); G *Paraharpes*, Ordovician (×3). (*After Hupé*)

that excessive ornamentation appeared in the trilobites at least 200 million years before the extinction of the group. Such youthful senility as this is a contradiction in terms.

It is much more likely that the variations seen in trilobites were adaptations to the varying environments in which they lived. The general broad and depressed structure of trilobites suggests that many of them were bottom dwellers. All the

THE EVOLUTION OF LIFE

organs are confined to the central lobe and the outer portions of the segments seem to have been protective rather than functional. The appendages were suitable both for crawling and for rather clumsy (as opposed to rapid and powerful) swimming. Other trilobites, however, were probably surface dwellers, especially some of those with thin 'shells' and many spines, which incidentally are almost always found in black shales associated with planktonic organisms suggestive of just this way of life. Other genera were probably burrowers, such as the trinucleids, for example, in which both the shape and the appendages suggest such a mode of life.

The brachiopods

The history of the articulate brachiopods is strikingly different from that of their inarticulate brothers. After appearing in small numbers in the Cambrian they became the dominant group in the Ordovician, where they were represented by no fewer than fourteen new 'super families'. They are characterized by a hinged calcareous shell, which is usually heavier and larger than that of the inarticulates. A number of different groups may be recognized (Fig. 12) showing almost every degree of variation in general form, ornament, ribbing, growth lines, and internal structure. The brachiopods are probably the most abundant and the most characteristic of all the Palaeozoic marine invertebrates, and their rapid evolutionary changes and widespread distribution also make them one of the most useful groups of index fossils.

The protozoans

The general absence of the *Phylum Protozoa* in the Cambrian probably implies that their earlier representatives lacked hard parts. By Ordovician times, however, a number of foraminifera certainly existed. This phylum is unique among animals in that all its members are organisms whose body consists of only a 'single cell', which performs all the essential life functions. Because, however, their whole mode of life is quite distinct from that of the metazoans (many-celled organisms) they are

THE HEY-DAY OF MARINE INVERTEBRATES

often referred to as 'non-cellular', rather than 'unicellular'. Typical representatives are *Amoeba*, and the *Foraminifera* and *Radiolaria*.

Most members of the phylum are microscopic in size, but exceptionally they may be several centimetres in length. The great pyramids of Gizeh in Egypt are built of a limestone which contains abundant large foraminifera (*Nummulites*), which early students erroneously interpreted either as coins or as lentils left from the food of the slaves who built the pyramids.

The body of protozoans is very variable in form, and they may be either solitary in habit or colonial. They inhabit a variety of moise environments (the seas, inland waters, soil, and the bodies of other organisms) and they move in a number of different ways. Some move by means of a whip-like 'tail' (flagella), others by means of irregular and temporary extensions (pseudopodia) of the protoplasm, so that they may be said to 'flow along', and still others by movements of numerous thread-like appendages (cilia) which are developed on the surface of the body. These organisms obtain food and reproduce in a number of different ways, some of them having elaborate reproductive cycles. Many of them are naked, but others, including the foraminifera and the radiolaria, build wonderfully intricate and beautiful skeletons (tests), usually of calcium carbonate or silica. These skeletons are so abundant in many marine environments that they form deposits of ooze that cover more than thirty-five per cent of the ocean floor.

For all their apparent simplicity, then, these unicellular organisms lead a 'full life' – they can move, capture and assimilate food, excrete waste products, secrete elaborate tests, respire, react to stimuli, and reproduce. Throughout most of geological time they have played and still do play an important part in the economy of nature. Some form the food supply of vast numbers of other small organisms, while others (the sporozoa) are parasitic disease-carriers, such as those which cause amoebic dysentery, African sleeping-sickness, and malaria. Some foraminifera were important rock builders during the later geological periods, and many are valuable

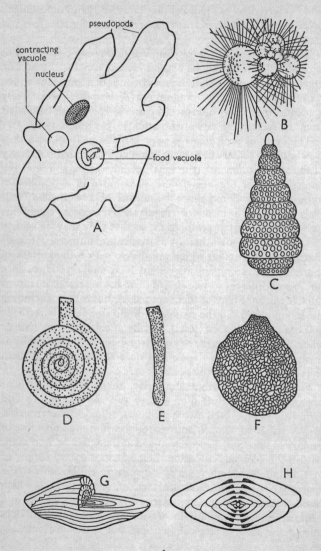

THE HEY-DAY OF MARINE INVERTEBRATES

index fossils in the determination of the age of the rocks in which they occur.

The protozoans, although they were almost certainly present, have not been definitely found in rocks older than the Ordovician, where they are represented by a variety of agglutinated foraminifera (Fig. 15) which assimilated a protective covering of sand grains.

More or less similar forms continue throughout the Palaeozoic and recent studies have shown that they may be useful index fossils in part of the Carboniferous.

The delicate and exquisitely sculptured radiolaria (Fig. 15C) are another group of minute protozoans. They are not common as fossils but they are known from Lower Palaeozoic times onwards. The remains of their living representatives are generally found only in the deeper parts of the ocean, and the rarity of such deposits in the fossil record probably accounts for the comparative rarity of radiolaria.

The corals

The newcomers to the Ordovician scene also included the corals, which subsequently became by far the most abundant fossil coelenterates. Many corals secrete a limy skeleton, some being solitary and living at great depths in the seas, but many others being colonial reef-builders. These are at present confined to shallow, clear, tropical waters where they often form conspicuous reefs. The Great Barrier Reef in Australia is over 1,200 miles long.

Though they are rare in the Ordovician, the corals became widespread in Silurian times. They were represented by a variety of forms, the calcareous skeletons of many of which are

---

FIG. 15 Protozoans.

A *Amoeba*, showing general structure; B *Globigerina*, a living foraminifera; C a radiolarian, *Lithocampe*, from the Devonian; D–H Palaeozoic foraminifera: D *Ammodiscus*, Silurian–Recent, E *Hyperammina*, Silurian–Recent, F *Saccammina*, Silurian–Recent, G–H a fusulinid from the Pennsylvanian–Permian, showing general structure; H is a view in thin section. (All highly magnified.)

THE EVOLUTION OF LIFE

so perfectly preserved as fossils that their detailed structure may be studied by means of thin sections. Many were reef builders and their remains are most common in thick limestone deposits such as those of Silurian age near Dudley, Worcestershire, and Chicago, within which the ancient reefs have been faithfully preserved. About 500 species are known from the Silurian and they include both solitary cup corals (Fig. 8) and a variety of colonial forms. One distinctive genus (*Goniophyllum*) was a solitary pyramidal shaped skeleton which was closed by a cover of four small plates.

Most of these forms represented the extinct group of rugose corals which underwent profound changes throughout the Palaeozoic. Several thousand species are known, some of which are world-wide in extent and are often of great value as index fossils (as for example in the thick Carboniferous limestone deposits of the Avon Gorge and those along the bluffs of the Mississippi). They are particularly conspicuous in the Devonian, both for their profusion and for their variety: some were more than two feet in length, and their delicate remains are preserved in many of the limestones on the South Devonshire coast (around Torquay, for example) and in many parts of North America (at the Falls of the Ohio, near Louisville, Kentucky, for example).

Almost all reef-building corals today have a very restricted occurrence, being limited to shallow, warm waters (mostly less than 150 feet in depth and above 64° F.). They occur mainly, therefore, in shallow tropical and subtropical waters (between about 28° N. and S. Latitude). Fossil coral reefs have a much wider distribution, and it has therefore been suggested that they may be used as an indicator of the temperature of ancient seas. It is not necessarily true, however, that these extinct corals had the same environmental requirements as their quite different modern descendants have, although they may not have been greatly different. Fossil coral reefs are known from Ordovician time onwards in latitudes that are now far from tropical (Scotland for example) and their lithology suggests relatively shallow water deposition. It may be nearer

THE HEY-DAY OF MARINE INVERTEBRATES

the mark to suggest that the presence of Silurian reefs, for example, in Ontario is indicative of a probably warm and more or less uniform climate in these areas, certainly more uniform than that of the present.

These ancient reefs (or bioherms as they are often, and perhaps better, called) are also of interest for quite a different reason, for they have recently been shown to be particularly favourable reservoir rocks for petroleum in certain areas (northern Illinois, for example).

Another group of coelenterates, which played a prominent part in the life of the Palaeozoic was the stromatoporoids, which deposited encrusting calcareous masses consisting of a series of successive layers separated by low pillars or columns. In many parts of the world they were important rock builders in Palaeozoic times.

The bryozoans

The bryozoans (*Phylum Bryozoa*) include a group of animals, many of which show a superficial resemblance to corals and colonial hydroids. They are, however, considerably more complicated, in that they have both a mouth and anus and a complete digestive tract. Most bryozoans are attached colonial marine organisms, which secrete chitinous or calcareous mat-like or frond-like skeletons (zoaria) a few inches in size (Fig. 16). These skeletons show great variety of form and are abundant as fossils. Each small animal is housed in a cup-like cavity (zooecium).

The bryozoans, though fairly common from Ordovician times onwards, are in many ways the Cinderellas of invertebrate fossils, for they have until comparatively recently been largely neglected by palaeontologists. Many of the earlier forms are rather massive cylindrical, dendritic, or encrusting types, but later types tend to be more slender branching and delicate lace-like forms. Most of these were only an inch or so in length, but in the Middle Devonian Hamilton formation of New York some forms reach a height of several feet.

Amongst the most problematical of all bryozoans are the

forms known as *Archimedes* (Fig. 16 E), consisting of lacy fan-like fronds arranged spirally around a solid screw-like axis. It has recently been suggested that this axis is not an intimate

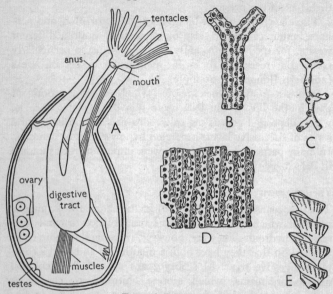

FIG. 16 Bryozoans.
A A living individual, showing structure (highly magnified); B–E Fossil bryozoa. The minute depressions each housed individual animals: B *Thamniscus*, Silurian–Permian (×10), C *Stomatopora*, Ordovician–Recent (×20), D *Fenestrellina*, Silurian–Permian (×3), E *Archimedes*, Carboniferous (natural size).

part of the animal but is, in fact, an algal growth, living in a symbiotic relationship with the lace-like fronds (the true bryozoan).

The pelecypods

In parts of the Ordovician gastropods are particularly common and they continue to be frequent but comparatively unimportant members of most later faunas. Ordovician times also saw

THE HEY-DAY OF MARINE INVERTEBRATES

the appearance of the pelecypods (*Pelecypoda*), which include a large group of dominantly marine bivalves, represented by such forms as the oyster and mussel. The two valves of the shell are usually alike in form and articulate along a hinge-line, being supported by both teeth and muscles. The body of the animal is laterally compressed, so that it may be completely enclosed by the shell. There is no well-developed head, but the foot is prominent and is variously adapted to different modes of life (Fig. 9 C).

Like the gastropods, the pelecypods, though fairly widespread as marine fossils from Ordovician times onwards, had a rather conservative Palaeozoic history and are rarely of major importance.

The cephalopods

Although the oldest cephalopods (Class *Cephalopoda*) appeared in the Cambrian, they did not become common until Ordovician times. They are the most highly developed class of molluscs, and include such forms as squids, the nautilus, and the octopus, as well as the extinct ammonites and belemnites. The cephalopods are exclusively marine, usually having a well-formed head (with eyes, beak-like jaws, and radula) surrounded by a ring of tentacles (hence the name 'head-foot') (Fig. 9 D). The animal may have two or four gills and may be naked (as in the octopus) or have an internal (as in the squids and cuttlefish) or an external shell (as in the nautilus). In contrast to most other molluscs which lead a sluggish existence, the cephalopods lead an active predatory life.

The shell is one of the most variable characters of the class, but the shell of *Nautilus* may be used to illustrate the major features of the external type shells. The animal lives in the open end of the shell in a living chamber into which it may withdraw itself. The shells are divided by transverse walls (septa) into other chambers, through which a fleshy stalk (the siphuncle) passes, perforating the septa and often being enclosed by septal necks. The lines which mark the junctions of the septa with the inner wall of the shell are known

THE EVOLUTION OF LIFE

as suture lines. These lines may be simple, angular, or extraordinarily complex (Fig. 9 J–M), these conditions being characteristic of the nautiloids (both living and fossil), the extinct goniatites, and the extinct ammonites respectively. The shells of fossil nautiloids and ammonoids vary greatly in form and size. The largest were more than fifteen feet long, some of the coiled forms were six feet in diameter, but most were a few inches in diameter.

The cephalopods appear first as a highly developed group in the Upper Cambrian. All the early Palaeozoic forms had simple nautiloid sutures (Fig. 9J), but they display a startling diversity of shell shape (Fig. 9). They are particularly common in the Ordovician, from which the largest known specimen (a straight-shelled type fifteen feet long) comes. These nautiloid forms continued to expand in the Silurian, but towards the close of this period the more complex ammonoid-sutured cephalopods developed from them. These forms, which are very valuable index fossils, displayed considerable diversity in suture and ornament during later Palaeozoic times, most of them being coiled forms.

The cephalopods are the most important of the Palaeozoic molluscs. Indeed their slender representation in modern seas (there are about 150 genera) would scarcely lead one to expect that during most of the Mesozoic era they were undisputed invertebrate champions of the seas.

The echinoderms

The remaining Middle Palaeozoic invertebrates were echinoderms, and they included the familiar starfish, sea urchins, sea cucumbers, and crinoids, as well as a number of extinct forms.

The starfish (*Asteroidea*) abound on most coasts. The body consists of a central disc, with five radiating arms, the mouth being located on the lower surface of the central disc, and the anus on the upper surface. (Fig. 11J). The lower surfaces of the arms each bear a central groove, bordered by double rows of tube feet. The whole body is encased in a leathery 'skin', studded by small plates and bearing on its upper surface a

THE HEY-DAY OF MARINE INVERTEBRATES

variety of spiny structures. Starfish have the ability to regenerate broken parts, and individuals are frequently seen with one or more newly formed arms. Starfish are savage predators of worms, crustaceans, and molluscs. They attack oysters by gripping the valves of the shell with their tube feet and slowly pulling them apart.

The *Ophiuroidea* (brittle stars) differ most conspicuously from true starfish in their small circular disc, with its five long, discrete, slender, jointed arms.

The *Echinoidea* (sea urchins, sand dollars, etc.) are enclosed in a more or less rounded, calcite test or shell, which is covered with movable spines. Some are discoidal and others heart-shaped. They may be thought of structurally as a starfish with upflexed arms meeting above the central disc, the intervening spaces being filled with calcite plates to form a rigid test (Fig. 11J). The test is usually more or less radially symmetrical and consists of twenty vertical columns of plates, arranged in ten series of double rows. The five of these corresponding in position to the arms of the starfish are perforated by the long tube feet, and the plates develop tubercles on which the spines articulate. The large mouth is situated on the lower surface and may be armed with five sharp teeth, while the anus is located on the upper surface of the shell. Echinoids are bottom dwellers, living on the sea floor, in tidal pools, or in burrows. They move by means of their tube feet and spines.

Starfish, brittle stars, sea cucumbers, and sea urchins are all free-moving forms. The remaining classes of the Echinodermata, are all fixed to the sea floor (at least during part of their lives) and are frequently grouped together as the *Pelmatozoa* (fixed or attached forms) in contrast to the *Eleutherozoa* (free or vagrant forms) which we have just discussed. The pelmatozoan body is generally encased in a hollow, globular structure known as the calyx. This is formed of a number of small calcareous plates and frequently carries the food grooves leading to the mouth on its upper surface. Its lower surface is usually anchored by means of a 'rooted' stem, composed of a large number of perforated disc-like plates.

THE EVOLUTION OF LIFE

The crinoids (*Crinoidea* – sea lilies, feather stars) are the only living representatives of the pelmatozoans. Many modern forms tend to live in relatively deep waters, and indeed the group was believed to be extinct until the close of the last century when the Challenger expedition took 10,000 individuals in a single haul! The animals have a rather flower-like appearance, the body being encased in a cup-shaped calyx of calcareous plates which bears five branching arms that assist in food gathering (Fig. 11 F). Some living and most fossil species were attached to the sea floor by a long, jointed stem, the disarticulated plates of which are common fossils and were used for beads by a number of primitive peoples (Fig. 11 I). The mouth and anus are situated on the upper surface of the calyx, though in some forms the anus is elevated on a long tube or proboscis.

Many modern crinoids are free-swimming (after an initial stage of attachment). The stalked species, often brilliantly coloured, are generally gregarious, and are so common as to form 'gardens' in some areas of the seas, often on coral reefs.

The blastoids (*Blastoidea*) are in a group of extinct bud-like fossils, whose small globular calyx is formed from thirteen plates arranged with exquisite symmetry. Five petal-shaped ambulacra (corresponding to the arms of a starfish) radiate from the mouth on the upper surface. The stem was either short or absent (Fig. 11 A–C).

The *Cystoidea* (cystoids) are another extinct group of variable form (Fig. 11 D). Both the body shape and the number and arrangement of plates in the calyx are variable, and only some forms developed stems.

We have already seen that the earliest echinoderms were the sea cucumbers, edrioasteroids, and eocrinoids. The cystoids became extinct in the Permian. Though locally abundant and varied in some Ordovician and Silurian strata, they were never an important group. Their exact evolutionary relationship is obscure: they were once regarded as the ancestors of all other stemmed echinoderms, but this now seems rather unlikely.

The other Cambrian echinoderms, the edrioasteroids, which

THE HEY-DAY OF MARINE INVERTEBRATES

are present as rare fossils until the Carboniferous, are regarded by some workers as ancestral to the echinoids. Though sea cucumbers appear in the Cambrian and both the starfish and echinoids in the Ordovician, they were comparatively unimportant throughout Palaeozoic times. Some of the ancient echinoids reached quite sizeable proportions, some of the globular Lower Carboniferous species being eight inches in diameter. The starfish, though common in some localities, are rarely important as fossils.

The sea cucumbers are too soft-bodied to be common as fossils, though their minute spicules are sometimes abundant in Middle Palaeozoic rocks. For the moment we may leave these groups: it was in the Mesozoic that the echinoids reached their acme and it is there that we shall meet them again.

The most important of the Palaeozoic echinoderms, however, were the crinoids and the extinct blastoids, both of which first appeared in the Ordovician. The crinoids were particularly common in many limestone deposits from Silurian times onwards, but the blastoids are comparatively rare until they reached a spectacular climax in the Lower Carboniferous. Perfectly preserved bud-like specimens occur in countless numbers in the limestones of the Mississippi River and parts of Europe (Fig. 11 B, C).

The arthropods

Amongst the crustaceans which appeared in the Ordovician were the ostracods – a group of microscopic, bivalved, more or less oval animals, which are still common in both marine and non-marine waters. They vary greatly in form, especially in the sculpture and ornamentation of the shell, and they occur in great abundance in many strata (Fig. 13 G–I). Many of the Palaeozoic genera display a prominent swelling on the shell, a feature entirely absent in living forms. This swelling provoked great speculation concerning its function, but it now seems clear that it is a brood pouch. These creatures, like trilobites, underwent a number of moults (usually eight to nine) and a statistical study of their valves indicates that the development of the swelling is a late feature of growth and is confined to only a portion (the females) of the population. Recent studies have shown that the pouches of some specimens preserved what seem to be larval forms.

Quite different from the ostracods are the extinct eurypterids, some of which reached a length of nine feet (Fig. 13 F), though most were much smaller (about one foot long). These ancient aquatic relatives of the scorpions had a rather blunted head with two pairs of eyes (one simple, one compound, made up of many facets) followed by a tapering body of thirteen segments with a long, spike-like tail. On the undersurface of the head region were six pairs of legs, the hind pair of which were often paddle-like; others were walking and balancing legs and some were developed into formidable pincers.

The whole body was covered with a thin, chitinous integument, which is often so well preserved that it may be extracted from the rock by chemical methods and retains its flexibility.

The oldest eurypterids come from the Ordovician: the group reached a peak of abundance in the Silurian and the Devonian and became extinct in the Permian. The presence of gills clearly shows the eurypterids to be aquatic, and the general structure of some suggests that they were powerful swimmers, though many were probably sluggish bottom dwellers. It is difficult to determine their habitat; they are not found in normal marine strata. They may have lived both in fresh-water lakes and streams and in brackish or saline lagoons and some have suggested that their environment changed during their history. They were probably carnivores, and may well have been the most savage predators of those early faunas.

Closely related to the eurypterids are the five living species of horseshoe crabs (including *Limulus*) whose earliest ancestors appeared in the Cambrian. The group as a whole has changed only little since Palaeozoic times, and *Limulus* is, in fact, often aptly spoken of as a 'living fossil' (Fig. 13 J). The forebears of *Limulus* were not, however, confined to the seas, for they are common in fresh-water strata of the Upper Carboni-

THE HEY-DAY OF MARINE INVERTEBRATES

ferous. One striking thing about the larvae of the king crab is their remarkable resemblance to trilobites.

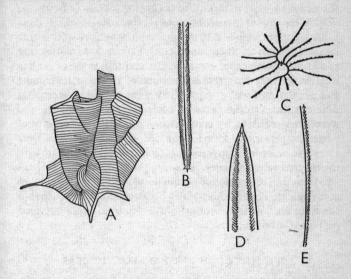

FIG. 17 Graptolites.

A *Diplograptus*, Middle Ordovician–Lower Silurian, highly magnified drawing, showing detailed structure of early growth stages (*after Wiman*); B *Diplograptus*, complete colony (natural size); C *Nemagraptus gracilis*, Ordovician; D *Didymograptus*, Ordovician; E *Monograptus*, Silurian. (C–E all natural size)

The graptolites

It is convenient to mention one further group at this point, for they form an intimate and important part of the ancient fauna of the seas. The graptolites are a group of extinct colonial marine organisms which occur in countless numbers in the dark shales of the Lower Palaeozoic. The earliest colonies, which appeared in the Upper Cambrian, were branching and dendritic in appearance, but later Ordovician and Silurian forms were marked by a decrease in the number of branches. Along the branches were developed a number of cup-like structures (thecae), each housing an individual organism, and these thecae vary widely in form. From a geological point of view the graptolites are of particular interest because of their great value as index fossils – they evolved rapidly and are world-wide in extent, most of them apparently having lived as floating plankton. Zoologically, however, the graptolites are of no less interest for they long represented one of the most tantalizing and perplexing fossil groups. They were variously referred at different times to the bryozoa, coelenterates, and plants, and were even regarded by some as inorganic. Recent studies of perfectly preserved forms extracted from limestones have shown, to everyone's surprise, that their closest affinities are with the invertebrate chordates. They are probably most closely related to the tiny colonial hemichordates (pterobranchs) of modern seas. The true graptolites became extinct at the close of the Silurian, but the more primitive dendroid types survived until the Carboniferous.

LIFE IN THE UPPER PALAEOZOIC SEAS

Lower Carboniferous times were marked by low-lying lands and widespread clear, shallow, limestone-depositing seas. In some areas (parts of Russia and Alaska) coal-forming swamps, lagoons, estuaries, and deltas already existed and these conditions spread throughout the Northern Hemisphere in Upper Carboniferous times. The warm, low-lying lands were bordered by clear, shallow seas, with a rich fauna of invertebrates and fish. In the Southern Hemisphere conditions were quite different. Many geologists believe that the southern continents (Africa, India, Australia, South America, and Antarctica) were parts of a single great land mass (Gondwanaland), and only later drifted apart. However that may be, these continents seem to have stood high above sea level, and there are late Palaeozoic glacial deposits in some areas.

New mountain chains were formed in many areas in Late Carboniferous and Permian times. The warm shallow seas

THE HEY-DAY OF MARINE INVERTEBRATES

persisted, and immense reefs developed in scattered areas (Durham and Texas, for example). Great thicknesses of salt and red sediments were deposited in land-locked basins and there was widespread glaciation in the Southern Hemisphere.

The marine life of this period was different in details, rather than in general character, from that of the Middle Palaeozoic. The protozoans, for example, were supplemented by one other group of great importance, the wheat-grain-like fusulinids, which persisted to the close of the Permian. These are amongst the most widespread of all foraminifera and their intricately sculptured shells are remarkably complex structures, especially as they are secreted by a single cell (Fig. 15 G, H).

These fusulinids, unlike most foraminifera, are sufficiently large to be clearly visible to the naked eye, and they have the added advantage that they may be best studied in thin sections of the limestones in which they occur.

Sponges, corals, and bryozoans continued with only relatively minor changes, but the trilobites declined to a handful of genera and many new groups of brachiopods, echinoderms, cephalopods, and other molluscs appeared. The overall faunal change was not, however, spectacular, for by this time all the major marine invertebrate groups were well established in their various ways of life, and the changes which they underwent were essentially minor in character, involving a closer adaptation to the environments in which they lived.

LAND, LAKES, AND RIVERS

So far we have been concerned with the great host of marine invertebrates that dominated the faunas of the Lower Palaeozoic, and as we have seen life was confined to the seas for the first 200 million or so years of the era.

The first undoubted record of the advent of life on the land is the primitive land plants from the Upper Silurian, about which we shall say more in the next chapter.

It is sometimes claimed that it was also in Silurian times that the first air-breathing animals appeared. The claim is based upon the presence of millipede-like creatures in certain Upper Silurian rocks, but the evidence remains rather doubtful. One thing, however, is quite clear. The complete colonization of the land by animals could only take place after the spread of fresh-water and land plants, although the earliest terrestrial vertebrates seem to have been carnivores. It is no accident that both land plants and land animals appear at roughly the same period of geological time. Some of the earliest terrestrial arthropods probably depended on the plants for food, just as some plants were later to depend on animals for pollination and still other animals were to depend on both for food. This interdependence of plants and animals is a reminder of the similar interdependence which exists throughout the world of living things. It is one of the major keys to an understanding of the evolutionary process.

The arthropods seem to have been the first animals to colonize the fresh-water environments. One ancient aquatic group which we have already discussed is the eurypterids, at least some of which seem to have inhabited fresh waters. Closely related to them, but differing chiefly in the presence of pincers on the second pair of limbs and 'combs' on the first trunk segment, are the oldest scorpions, found in the Silurian of both Europe and North America. These are thought by some to be the oldest air-breathers, but it seems more likely that they lived in inland water. By Carboniferous times, however, descendants of these forms were clearly air-breathers, and the Carboniferous also saw the appearance of the first true spiders, although both spider-like and tick-like creatures have been recorded from the Middle Devonian Rhynie Cherts of Scotland. The insects were already well established by Upper Carboniferous times, and we shall consider them in Chapter 10.

It is worth pausing to note the significance of these arthropod successes in the conquest of the land – by Devonian times ticks and spider-like creatures, by Carboniferous times true spiders and such insects as scorpions, cockroaches, and primitive dragonflies, had all established themselves firmly ashore. The

THE HEY-DAY OF MARINE INVERTEBRATES

extent of their success can best be measured by comparing them with other groups, for the ability to live on land is an uncommon one amongst other animals. Apart from the vertebrates, only worms and gastropods are common air-breathing animals. In the arthropods, however, the habit was taken over

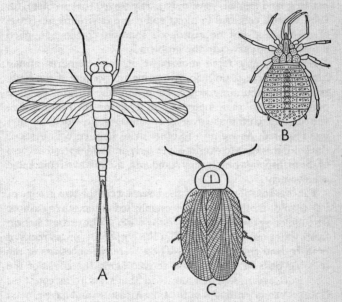

FIG. 18 Carboniferous arthropods.
A *Eubleptus*, a primitive winged insect; B *Eophrynus*, a spider; C *Aphthoroblattina*, a cockroach (*after Handlirsch*) (about natural size).

with spectacular success by every large order. Some indeed are so well adapted that they are equally at home in air or water – land crabs, for example. Now how is it that the arthropods should have been so successful in a competition in which other groups have achieved so much less? The answer seems to lie in a number of peculiar characteristics, each of which gave them a flying start. Two of these are of particular importance. Firstly,

THE EVOLUTION OF LIFE

the arthropods were well protected by their chitinous covering against the hazards of desiccation which land life involves. The body functions depend upon a constant internal supply of fluid. In water there is little danger of any moisture loss except through osmosis, but on land evaporation is continuous. Some existing land animals have only partly solved this problem, for they are still confined to moist and damp environments (frogs are an example) but the arthropods' chitinous envelope supplied a ready-made answer to the problem.

Secondly, any rapid movement on land requires strong limbs, for there is no help in bearing the weight of the body from the surrounding water. Here, too, the arthropods were pre-adapted to their new way of life, for their numerous, varied, and chitinous-covered appendages provided all that was required. As we turn to other more familiar conquests of the land, it is easy to overlook the early and widespread success of the arthropods but, by any standards, theirs was a remarkable achievement.

Far less conspicuous than the insects were the two groups of Palaeozoic molluscs which established themselves ashore (though one group still clung to the water). The earliest known fresh-water mussels are found in the Upper Devonian rocks of both Ireland and New York. These were the ancestors of the teeming beds of mussels that thrived along the edges of the later coal swamps. Their crowded shells occur in countless 'mussel bands' throughout the Coal Measures and have proved to be of the greatest value in the correlation of coal seams over wide areas.

The oldest land snails are known from rocks of Upper Carboniferous age in Nova Scotia where they are found inside the fossilized stumps of ancient trees. In some later periods freshwater snails became so common that their remains make up most of the rock in which they occur. The beautiful Jurassic Purbeck Marble, used extensively in interior church decoration, is crowded with the remains of one genus, *Viviparus*.

It is well now perhaps to glance quickly back, for we have covered about 375 million years of invertebrate history in the

THE HEY-DAY OF MARINE INVERTEBRATES

course of these few pages. As the long Palaeozoic era drew to a close virtually all the familiar groups were already well established. Many others lived with them. But the end of the Palaeozoic times was, in one sense, the end of the dominance of the invertebrates, for during those long ages they were dominant, yet not alone. Across the once barren earth that saw their advent, a mantle of green was slowly spreading. In the swiftly flowing streams and in the restless seas other forms of life were taking their place. Wood and bone – these were the commodities which were to alter the whole history of life on the land. With their advent, a new age dawned.

Chapter 6

ANCIENT PLANTS

PLANTS to most of us are flowers, bushes, trees, and, usually as an afterthought, grass and moss. But there are plants and plants, and although almost all of them share the basic property of photosynthesis (the manufacture of carbohydrates from water and carbon dioxide by the action of sunlight on chlorophyll), they differ from one another in other ways no less striking than the ways in which amoeba differs from man.

The thallophytes

The plants fall into three major divisions. The first, the Thallophyta, includes the most primitive. These are either unicellular or consist of loosely organized groups of cells, and most of them inhabit damp or aquatic environments. They lack many of the features that we tend to regard as typical of the plants; such things, for example, as stems, roots, leaves, or woody tissue. But if they are so un-plant-like, what qualifications have they to be considered as plants at all? Some will suggest that the answer may be found in one word – 'chlorophyll'. In spite of all appearances, they 'live' as plants.

But this simple answer does not solve our problem of recognizing a plant – if anything, it makes it more confusing, for instead of providing a convenient clear-cut dividing line between plants and animals it points us to the hazy overlapping zone between the two kingdoms. And just as the viruses carried us back to the threshold of life, so these lowly thallophytes carry us to the ill-defined threshold that separates the plant world from the animal.

Now many of the protozoans are, as we have seen, clearly animals – they move, grow, assimilate food, and excrete waste products very much as 'undoubted' animals do. But there are some tantalizing exceptions. Let us look for a moment at the tiny unicellular organism *Euglena*, a common inhabitant of ponds

ANCIENT PLANTS

and ditches. It has a more or less oval body which is moved through the water by movements of the flagellum; the creature can also crawl and perform worm-like movement: in other words it is capable of typically 'animal' movement – but it contains chlorophyll and obtains nutrition by photosynthesis!

Euglena is really a living contradiction to most of our ideas about the differences between animals and plants, and the contradiction arises, not because we can't decide which of the two it is, but because it appears to be both. Other forms which are very closely related lack chlorophyll and behave as any other animal, using the long thread-like lash to swim, taking in and digesting food, and so on. The implication of this is clear. 'Plants' and 'animals' are abstract categories of our own making – conceived and formulated purely as a matter of convenience. Because of this, it by no means follows that all organisms must fit into one group or the other. Perhaps *Euglena* is a living remnant of the ancient and primitive group of minute aquatic organisms which were the ancestors of both animals and plants. But can we not resolve the conflict by considering chlorophyll as distinctive? Can we suppose that 'if chlorophyll – then a plant' will give us a safe rule? Unfortunately this too will not do, for some of these thallophytes (the fungi) which in other respects are very plant-like, do not possess chlorophyll. In fact, these fungi represent a problem family – for in various members within it, almost all the 'typical' plant characters (need for sunlight, absence of movement, and so on) break down. And yet, on balance, its members seem to be plants.

Now what are they like, these perplexing thallophytes? They contain three broad classes, the Fungi, the Algae, and the Bacteria. The Fungi include the moulds, mushrooms, and lichens. The moulds exist as scavengers of the plant world – they feed upon either living or decaying matter. They contain no chlorophyll, and their food-gathering is more reminiscent of animals than of plants; for they obtain nourishment by attacking the substance on which they grow, secreting digestive juices and then reabsorbing them, charged with dissolved substances. The fungi are also unusual in their structure, for they

THE EVOLUTION OF LIFE

are not multicellular, but consist of a flowing mass of protoplasm – with numerous nuclei, flowing within a thread-like covering (the hypha). Fungi reproduce asexually by means of minute air-borne spores, but in many sexual reproduction takes place within a complex cycle. The spore-bearing structures are often much more conspicuous than the rest of the plant: the mushrooms and toadstools are good examples. In spite of the apparently lowly nature of the fungi, they play an important part both in the economy of nature (where they are important agents of decay) and in the life of mankind. Some are a scourge to crops, a few produce disease in man (e.g. athlete's foot), and still others are becoming increasingly useful as the source of antibiotics (penicillin, streptomycin, etc.).

The yeasts and lichens (the yellow-green growths that encrust trees and old walls) are also included within the fungi, although the lichens are actually a symbiotic association of both fungi and algae. Because of their soft structure they are not common as fossils, but their spores may be preserved and some fossil plants show evidence of the action of fungi.

The Algae include the seaweeds, the green scum of ponds, the exquisitely coloured organisms of hot springs, and the minute and delicate diatoms, myriads of whose delicate siliceous shells are found throughout the waters of the world. Many of the seaweeds grow to a very large size, the largest being more than 200 feet in length, but most of them leave little or no fossil trace, though many indistinct carbonaceous markings are often attributed to them. There are, however, some important exceptions, and amongst these are the calcareous algae, which are found in both fresh-water and marine environments. These organisms trap thin layers of calcium carbonate over successive layers of cells, and so in time build up layered deposits, which reflect some of the cellular structure. These limy structures may be tube-like in shape (as in some of the green seaweeds) or spherical. In some waters algae are so common that their calcareous deposits become important rock builders, and this has also been the case in the geological past (Fig. 19E). We have already seen that some of

ANCIENT PLANTS

the earliest fossils known are calcareous algae, and their remains make up a large proportion of some strata of more recent age, such, for example, as the Dolomites of the Alps. The charophytes are small, globular, spirally striated, fruitlike structures which are found as fossils in some fresh-water sediments. They are derived from a more advanced type of algae, which have stems and leaves (Fig. 19D).

The diatoms are much less conspicuous than the calcareous algae, but they occur in such teeming numbers that they too have been and are important rock builders in some areas. They are minute, unicellular organisms, which secrete a delicate siliceous skeleton, and which are found both in the sea and in fresh water. They often flourish in areas where other life is relatively sparse, and the silica of their skeletons is unusually resistant to solution after their death (Fig. 19F). These skeletons form a deposit which is known as diatom ooze and which is also found consolidated as a rock (diatomite) in some parts of the geological column. We have already (p. 86) said something of the earliest fossil remains of algae. They appear to have existed in the earliest seas and they have continued to provide the essential grass of the oceans and basic link in the complex food chain of the seas.

The Bacteria are minute unicellular plants of diverse shape and form which have no hard parts, but which live in a great variety of environments, including other organisms. Some achieve a limited amount of movement by means of flagella-like structures. They lack chlorophyll, have no 'typical' nucleus (although there is a type of nucleus present), and reproduce mainly by means of simple transverse fission. In spite of their lowly character, they play an important role in the life of both animals and plants. Some are benefactors on which the familiar activity of life depends (for example in decay and nitrogen fixation), but others are agents of disease and death to both animals and plants. Their lack of hard parts and minute size makes them uncommon and inconspicuous fossils, but there are a few records of their former presence, both in very ancient rocks (p. 86) and also in some fossil animals, as in

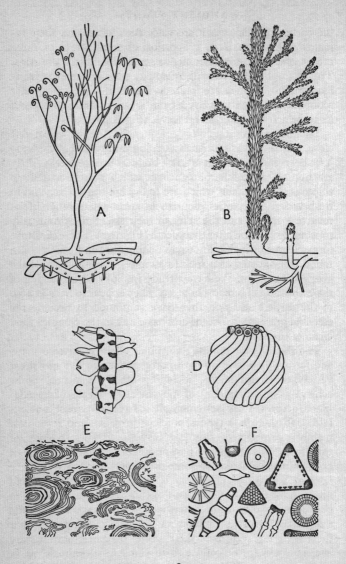

ANCIENT PLANTS

the armour of a Devonian fossil-fish. Some geologists maintain that the vast iron ore deposits of many parts of the world owe their origin to the iron-secreting action of bacteria.

We have already seen that remains of a number of these lowly thallophytes extend far back into the Pre-Cambrian. They are at present, in fact, the oldest known organisms and (though predictions are rash) they will probably retain that distinction.

The bryophytes

The second great division of the plant world is represented by the Bryophyta, which include the familiar mosses and liverworts of damp, shady places. They resemble the thallophyta in their lack of true roots and vascular (water carrying and supporting) tissue, but they represent a marked advance over them in their possession of leaves and stems. They reproduce by means of spores and there is an alternation of sexual and asexual generations. The sexual generation is the larger of the two, and the asexual spores are attached to it. The lack of any woody tissue makes the Bryophyta rare as fossils, and very few have been described. The oldest representatives of both mosses and liverworts are found in the Carboniferous (Fig. 19 C).

The tracheophytes

The third and by far the largest division of the plant kingdom is the Tracheophyta, the vascular plants, the only major group of plants fully adapted to terrestrial life. The division includes a great variety of forms, but all share in the possession of

---

FIG. 19 Palaeozoic plants.

A–B Devonian land plants: A *Psilophyton*, height 2–3 feet (*after Dawson*), B *Asteroxylon* ($\times \frac{1}{2}$) (*after Kidston and Lang*); C Carboniferous bryophyte *Hepaticites*, from the Coal Measures (approx. $\times$ 10) (*after Walton*); D fossil *Charaphytes*, the spores of algae (greatly magnified); E calcareous algae *Cryptozoan* from the Cambrian, length of rock surface shown is about 6 feet; F variety in living diatoms (approx. $\times$ 50).

THE EVOLUTION OF LIFE

vascular tissue, which carries water and food and also supports the woody structure of the characteristic stem, leaves, and root. This woody structure makes the tracheophytes susceptible to fossilization, and they have a long fossil record.

Although spores are known from rocks of Pre-Cambrian and Cambrian age in Siberia, the oldest known fragments of land plants have recently been described from the Ordovician of Poland and the Appalachians, but the oldest well-preserved land plants come from the Silurian of Victoria, Australia. The flora (which had floated out to sea and is associated with a typical Silurian graptolite *Monograptus*) includes two very primitive, but nevertheless quite distinct, types of plant. One genus, *Baragwanathia*, resembles the later Lycopods (p. 152) and bore the spore case (the *sporangium*) near the upper surface of the leaf. The plant had branching stems up to about two inches in thickness, bearing slender, close-set leaves. The other group, which is represented by the genus *Yarravia*, became world-wide in extent during later Devonian times. A number of erect cylindrical and leafless stems rose from a tangled basal mass of branches. These stems forked and bore spore cases on their tips. The whole plant was small, most of the branches being only about $\frac{1}{10}$ inch in diameter. Well-preserved spores of both these genera are known. The plant is clearly very primitive – it has no differentiated stem, leaf, or root systems for example, but it does exhibit the essential features of all later land plants – the presence of a vascular system and cuticle.

By Devonian times there are widespread occurrences of fossil land plants. Well-preserved floras are known from the Lower Devonian of Wyoming, the Gaspé peninsula, Belgium, Norway, Britain, and Spitzbergen, and the Middle Devonian of Scotland, Bohemia, and China. All these plants bear a broad resemblance to one or other of the two broad types we have just discussed (see Fig. 19A, B).

Perhaps the best known of these floras is that from the Middle Devonian Rhynie Chert of the Muir of Rhynie, Aberdeenshire. The silicified plants occur in their positions of growth, and are so perfectly preserved that even the nuclei of some of the cells may be identified. Four species are repre-

ANCIENT PLANTS

sented, all very simple and rather small forms (the largest being perhaps eighteen inches high). With these plants is an equally well preserved fauna of eighteen species of primitive spider-like creatures, a form of tick, and a wingless insect.

This remarkable assemblage serves to emphasize again the intricacy of evolutionary change. It is no mere chance that Devonian times marked not only the spread of land plants, but also the establishment of terrestrial arthropods and vertebrates. It was only when vegetation was well developed that non-carnivorous land animals could have a source of food, just as it was only later when certain insect changes were taking place that dependent types of plants developed with them.

These primitive vascular plants are usually assigned to a group known as the *Psilophyta* or psilopsids, of which only two genera now survive. In comparison with most living plants they were very simple, having vascular tissue but no seeds, probably lacking roots, and those that had leaves at all having small insignificant structures. Indeed for what they were they were a small and unpromising-looking group. But if palaeontology has a moral it is that what an organism is, is often of less importance than what it may become. So it was with these first lowly land plants; for all their insignificance they were the forerunners of the mighty Coal Measure forests.

But in fact not all Devonian land plants were quite so small as these. In New York State, for example, there have been discovered the remains of a forest composed of vegetation not unlike that of the Coal Measures. The trees, some of which had a diameter of two feet, included representatives of a number of groups which dominated the coal-forming swamps. It is to these that we must now turn.

THE COAL SWAMPS

We need first to see the wood before the trees, to catch a glimpse of the general character of these coal swamps. Perhaps the most striking thing about them was their extent – they spread from Britain across western Europe, and eastern North America. The presence of numerous marine bands in Coal Measure strata shows that most of the swamps were near sea level: coastal paralic basins, perhaps not unlike the coastal, tide-level swamps of the Gulf Coast, the great estuarine lowland swamps of India, and parts of the Far East. The fossil trees themselves have features (such as the smoothness and thickness of the bark, their prolific growth and luxuriant foliage, the absence of seasonal growth rings, the large size and thin walls of many of the cells, and so on) which suggest that the peaty soil in which they grew was almost permanently waterlogged. The climate was apparently warm or hot and humid, with abundant and perennial rainfall. It was the rain which maintained the swamp conditions, and these, in turn, that prevented the rapid decay of the plants and so led to the formation of coal. Under such conditions in existing forests, trees grow very rapidly – often ten feet in a year. In such spreading tangled forests, the great trees of the coal age flourished, trees that are quite unfamiliar to our modern eyes. The flora of the inland (limnic) coal-forming basins, was less diversified, although not greatly different in general character.

The *Sphenopsida* (arthrophytes) include the scouring rushes or horsetails, which have simple, hollow, longitudinally ribbed, jointed stems, from the nodes of which circlets of leaf-whorls grow. They reproduce by means of spores, which are borne in cones at the ends of the stems. Most modern members of this group are rather small but in the great coal-forming swamps giant arthrophytes reached a height of seventy feet or more. They grew in imposing strands and bore long graceful leaves (known as *Annularia*) which were unlike the stubbier leaves of living forms. In spite of their great size, their trunks were pithy, rather than woody, but they were supported by thick bark (Fig. 20C, 1).

The *Lepidophyta* or *Lycopods* include the modest club mosses and ground pines amongst their living members. These are small, trailing, herbaceous plants, which bear spores in terminal cones. They have true roots, leaves, and stems con-

ANCIENT PLANTS

sisting of woody tissue and pith, but the stems are jointed. Like the horsetails, they are today a much-reduced group, their forerunners in the Carboniferous coal swamps being the giant scale trees. In these, short strap-like leaves were attached to the stem, on which they left a scale-like scar which shed. These mighty trees, often with a diameter of six feet and reaching a height of more than one hundred feet, are amongst the most important constituents of coal.

Two genera are particularly common. *Lepidodendron* (Fig. 20 A, F) had a tall slender trunk, which branched repeatedly near the top. These upper branches were covered with slender pointed leaves and the spore cases were borne on the tips of the limbs. The diamond-shaped leaf scars were arranged in spiral rows. *Sigillaria* (Fig. 20 B, G) had an altogether different appearance: a stouter, usually unbranched trunk, whose upper portion bore a brush-like fuzz of longer blade-like leaves, scars of which were arranged in vertical rows separated by ribs on the bark. More than a hundred species of each of these two genera are known, though not all were of such gigantic proportions. The typical roots (*Stigmaria*) of the scale trees (Fig. 20 E) were massive spreading structures.

The *Filicineae* are the true ferns, one of the simplest of the groups of plants, which, although adapted to life on the land, are still dependent on damp conditions. Ferns are familiar members of most of our woodlands and many moorland areas, but tropical ferns are much larger than these, often reaching fifty feet in height, and the fossil ferns found in such rocks as those of the Coal Measures were sometimes equally spectacular. Ferns reproduce by means of spores, which are borne either on the underside of the leaves or on specially modified leaves (Fig. 20 D).

The seed-bearing plants

Now all the plants we have considered so far are somewhat imperfectly adapted to life on the land, for the critical stages of their reproduction are dependent on the presence of water. The sperms are minute flagellated cells, which must swim through

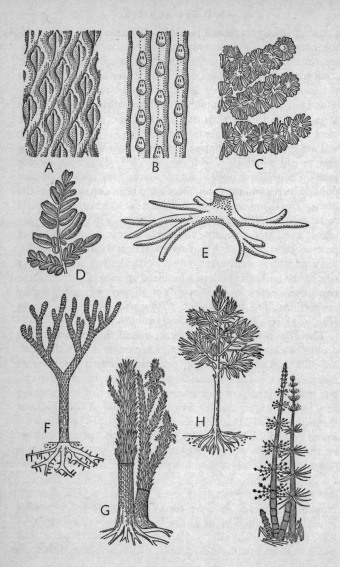

ANCIENT PLANTS

a liquid in order to fertilize the eggs. Because of this dependence on water many ferns and all the early plants were restricted to damp lowlands. The seed ferns (*Pteridospermophyta*), however, and all those that follow are true seed-bearing plants. In them the male microspore is represented by pollen grains and the female megaspore is retained and fertilized within the protective and nutritive structures known as seeds. This gives the newly fertilized 'plants' a much better chance of survival than that afforded by free spore-bearing plants and so enables the plants to survive even in arid conditions. The development of seeds was clearly a major step in the adaptation of plants to life on the land, although in the earliest plants of this kind the seeds were not completely enclosed and protected, as they are in modern flowering plants.

The Pteridospermophyta flourished during the Upper Palaeozoic, and they probably gave rise to the later seed-bearing plants. In general appearance they are very like the true ferns, from which they probably arose, and they differ from them chiefly in their method of reproduction – they bore small nut-like seeds rather than spores.

The *Coniferophyta* include the great variety of living and fossil conifers. Such trees as the pine, spruce, larch, cedar, monkey puzzler (*Araucaria*), sequoia, and cypress are familiar examples. Most members of the group have an evergreen foliage, which consists of needle or strap-like leaves, and the naked seeds are generally borne in cones. Many of these forms have a long fossil history. One primitive group (the Cordaites)[1]

1. Some botanists regard the Cordaites as a distinct group.

FIG. 20 Coal Measure plants.
A–B Lycopods: A *Lepidodendron*, showing leaf scars in oblique rows, B *Sigillaria*, showing leaf impressions in vertical rows; C *Annularia*, the foliage of scouring rushes; D *Neuropteris*, a fern-like plant; E '*Stigmaria*', the fossil roots of lycopod trees; F–I restorations of Coal Measure plants (*after Dunbar*): F *Lepidodendron*, G *Sigillaria*, H *Cordaites*, I *Calamites*.

were amongst the mightiest trees of the Coal Measure swamps, and towards the close of the Palaeozoic era the conifers emerged as the dominant floral group (Fig. 20 H).

CLIMATE AND DISTRIBUTION

We have already seen that, in Devonian times, floras were world-wide and uniform – that they exhibited no obvious regional variation. This implies an absence of sharply zoned climates, with the present polar and tropical extremes subdued. By Upper Carboniferous times, however, things were quite different. The great coal forest trees which we have described were characteristic of western Europe and eastern North America. The same flora extended into North Africa, but its eastward extension into South-East Asia (and possibly western North America) was represented by a rather different flora, the Cathaysian, dominated by the plant *Gigantopteris*.

To the north and south of this central floral belt two quite distinct floral zones were present, the Angaran in northern Asia and north-eastern Russia, and the Gondwanan, characterized by *Glossopteris*, in the Southern Hemisphere. The broad similarities between these two floras are probably the result of similar ecological adaptation, rather than genetic affinity, and the form of their plants suggests growth in temperate conditions, the growth rings implying seasonal changes. This is in striking contrast to the more uniform warm or subtropical forests of the central European–American–Cathaysian zones. Indeed, the southern Gondwana province showed much less diversity than the northern Angaran flora, and this lack of variety is also shared by its insect fauna. This may well be a reflection of the harsh glacial conditions which had already begun in the southern continents.

The origin of these well-marked floral zones presents some tantalizing problems, for we have to account not only for the changes within the plants themselves, but also for the physical conditions which accompanied (and perhaps promoted) them. It is not difficult to understand how isolation and adaptation

ANCIENT PLANTS

to varied conditions could have brought about the floral changes. But it is difficult to understand the apparent physical transition from the uniform climatic conditions of Devonian time to the restricted and zoned climates of the Upper Carboniferous. It may be that this marks a major change in the conditions of the earth's surface – or it may be that we have misread the implications of the floras. Perhaps the floras of Devonian times were sufficiently tolerant to flourish in various climatic environments. Whichever is the case, it remains a major problem.

Such were the mighty forests through which the sunlight of distant Carboniferous skies filtered. Sultry, perpetual swamps, with gigantic insects creeping through the tangled undergrowth and filling the heavy air. They were not to survive for long, but before we follow their decline, we must examine one final aspect of life on the land – and to trace its origins we return for a while to the water.

Chapter 7
THE RISE OF THE VERTEBRATES

VERTEBRATE ORIGINS

We have so far seen two major developments in the life of the Palaeozoic – the dominance of the marine invertebrates with the subsequent adaptation of some forms to life on the land and the development of terrestrial vegetation. There is, however, one more event to be considered – an event of peculiar interest because we, who live 400 million years after, are its products.

We assign ourselves to the chordates (*Phylum Chordata*), a group which includes a startling variety of animals, ranging from the lowly tongue worms to the host of such familiar and conspicuous vertebrates as mammals, birds, frogs, and fish. Most representatives of the group are free-living forms, and they inhabit almost all environments.

The affinity of these apparently radically different groups is indicated by three distinctive features which their embryos share – an axial supporting 'rod', the notochord, which in the higher forms is replaced by the vertebral column, a single dorsal nerve tube, and gill slits. These structures may be modified or lost in later life. Most chordates also have bilaterally symmetrical segmented bodies, with complete and relatively complex digestive tracts.

The more inconspicuous members of the phylum are distinguished as the *Acraniata* and lack a brain and skull. They are typically small, marine creatures, whose most common fossil representatives we have already met in the extinct graptolites (p. 137). Other fossil representatives are rare, but acorn or tongue worms, sea squirts, and lancelets are living members of the group.

The remaining chordates which are commonly found as fossils are the vertebrates: all have an enlarged brain, enclosed

THE RISE OF THE VERTEBRATES

in a skull (cranium – hence the term *Craniata*, by which they are described). The body is supported by a segmented vertebral column and they show many specialized features. They constitute eight classes, half of which are 'fish' (*Pisces*) and half tetrapods.

Before we look at them in detail, let us say a word about their origin. Origins are one of the most important but also one of the most puzzling aspects of palaeontology. We have already seen this with the Cambrian fauna, which appears with Melchisadechian abruptness, fully grown, without any obvious ancestors. The same is true of most of the major groups of organisms (see p. 280) and the vertebrates are no exception, though fortunately there are good fossil links between some of the various classes. The oldest vertebrates appear in the Lower Ordovician of Russia and the Middle Ordovician of the western United States (Wyoming, Colorado, and South Dakota), where they are represented by fragments of the bony armour of fishes. We know almost nothing of the fishes themselves (they do not appear again in any numbers until the Middle Silurian), but the presence of bone clearly marks them as vertebrates and distinguishes them from all other groups. This then is the sum of the direct fossil evidence concerning the origin of vertebrates – bone suddenly appears in the Ordovician. The most obvious and fruitful way of tackling the question of their origin is that of comparative anatomy. Let us therefore look at the invertebrates and ask which, if any, group might provide an answer to the problem of vertebrate ancestry.

A backbone either developed around or else replacing a notochord, a central nervous system *above* it (dorsal in position), an alimentary tract or gut *below* it (ventral in position), and gill slits – these are the characters which all vertebrates possess at some time or another of their lives.

Now our immediate tendency is to cast around for other organisms from which these structures may be developed, and this is clearly a sensible starting point. There are just two points to be borne in mind, however. Firstly in the search for origins we can expect to obtain more clues by studying the

THE EVOLUTION OF LIFE

simplest chordates than the more complex: and secondly the embryonic development of any creature is often quite unlike the adult. This latter point is of particular importance, for the results of many quite independent studies demonstrate that

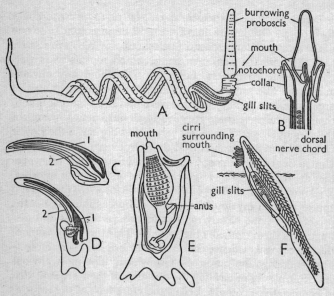

FIG. 21 Primitive chordates.
A–B *Balanoglossus*, the tongue worm (a hemichordate): A external appearance, B detail of head; C–E tunicate or sea squirt *Ciona* (a urochordate): C younger and D older larva of tunicate metamorphosing into sedentary adult E (1 dorsal nerve cord, 2 notochord); F *Amphioxus* (a cephalochordate), to show similarity to tunicate larva above.
(*After Simpson, Pittendrigh, and Tiffany.*)

the larval or embryonic form of an organism often provides the most valuable clues to its origin and affinities. The tunicates or sea squirts are a good example of this (Fig. 21 C–E). These are sedentary, sac-shaped, almost sponge-like creatures, encased in a stiffish 'tunic', most of which are attached,

THE RISE OF THE VERTEBRATES

sometimes in colonies, to rocks and seaweeds in shallow seas.

Anything less like a vertebrate is hard to imagine – for although they have gill slits the sea squirts have no nerve cord, no vertebrae, and not even a notochord. But the larval form is quite different – it is free-swimming, with a tadpole-like body which has both a notochord and a nerve cord above it (Fig. 21C).

Now the larval sea squirt is remarkably like another primitive chordate, *Amphioxus* the lancelet, which inhabits many coastal waters (Fig. 21F). This streamlined translucent animal, usually about a couple of inches long, spends much of its time with all but its anterior end buried in the sand, although it swims adequately. It has a notochord, a dorsal nerve cord, a simple ventral digestive tract, and gills.

There is therefore a striking similarity between the adult *Amphioxus* and the larval tunicate. This at least is a help, for such simple and unspecialized chordates, although they have certainly undergone some modification, are probably not too greatly different from the earliest members of the group. With no real head or brain, no organs of sense, no backbone, and no paired fins, this lowly animal is a far cry from the salmon, the eagle, and the lion – yet it may not be too unlike the primitive ancestors from which they all arose.

There is, however, one chordate group that is even more primitive than these two. The acorn worms are superficially far more similar to typical invertebrates than vertebrates (Fig. 21 A, B). They are sluggish, soft, worm-like animals that live in burrows in the sands and muds of shallow marine waters. They feed by extending the mouth from their burrow, to pass through their bodies a steady stream of mud, from which they extract the organic matter. The mouth is usually somewhat hidden by the proboscis, a conspicuous, muscular, cylindrical or conical burrowing structure, that retracts into a sheath-like collar (and resembles an acorn).

Now, in spite of appearances, these animals have distinctive chordate structures: a dorsal nerve cord (as well as a shorter ventral one), gill slits, and a notochord. It is, however, their larvae that are of particular interest. These are small, floating, bell-like organisms that have a twisted ring of cilia surrounding the mouth. They bear some resemblance to the trochophore larvae of worms and molluscs, but they differ in that these latter have only a single ciliated ring which encircles the body in front of the mouth. It requires no great stretch of the imagination to see that such simple creatures as these larvae might well be similar to the forms that gave rise to the later vertebrates. But we have still not traced the story back very far. Can we find any further clues which would suggest not just the origin of the vertebrates but the origin of the chordates?

There have been many suggestions of the route by which the chordates may have evolved. It was once thought, for example, that the arthropods or annelids provided the most promising ancestors, but it now seems unlikely that their dorsal gut and ventral nerve cord could become completely inverted to produce the quite opposite characteristic chordate arrangement. The most likely ancestors of the chordates are now believed to be the echinoderms. This is something of a surprise, for the adult echinoderm bears precious little resemblance to any known chordate. Here again, however, appearances are deceptive, for the larval echinoderm is remarkably similar to that of the acorn worms; as a matter of fact the larva of *Balanoglossus* was mistaken for an echinoderm when it was first discovered.

Nor is the resemblance confined only to external appearance, for there are other important physiological, anatomical, and biochemical similarities between them. This does not prove the echinoderms to be the ancestors of the chordates, but it does suggest that both the echinoderms and the chordates may have been derived far back in geological time from a common ancestor – now extinct and otherwise completely unknown.

It was perhaps by some such route as this that the vertebrates developed. We shall probably never know, however, exactly what the first chordates were like, for they were almost certainly soft-bodied and therefore unlikely to be preserved as fossils. It seems very likely that the heavy bony armour of the oldest fossil vertebrates is, in fact, a specialized outcome of a long geological history. The reason for the development of

THE RISE OF THE VERTEBRATES

armour is far from clear, but it has been suggested that it may have been a method of defence against contemporary predators.

The oldest fossil vertebrates can tell us little more than this. Their remains are so rare and so fragmentary that we know almost nothing of the animals they represent. Vertebrates remain rare through the Ordovician and most of the Silurian: it is only in the Devonian that they become fairly common.

LIFE IN THE WATER

All four of the classes of 'fish' have gills, skin with scales, and fins (which are paired in the higher forms). The most primitive are the *Agnatha*, which are the only craniates lacking true jaws and paired fins. This group includes such living forms as the lamprey and hag-fish as well as the extinct ostracoderms, which are the oldest known vertebrates.

The Agnatha

These primitive agnatha show a good deal of variation, which can be illustrated by considering two extreme genera. *Jamoytius* is a small, fish-like creature from the Upper Silurian of Scotland. The torpedo-shaped body bears both median and lateral fin folds, and has large eyes: superficially it is not too unlike *Amphioxus*. It occurs in estuarine shale. *Jamoytius* is, however, not typical of the early fish, almost all of which belong to the heavily armoured group known as the ostracoderms. The entire body of some ostracoderms, such as *Thelodus* (Fig. 22B), was covered by a very large number of bony 'scales' but most other forms have a heavy bony head shield, with a trunk and tail region protected by bony scales. *Cephalaspis* (Fig. 22E,F), which reached a foot in length and is found in the Devonian, was typical of the group. *Cephalaspis* lacked many characteristics of later vertebrates, having neither jaws, internal vertebral column, nor true paired fins. It has a massive bony head shield, with closely set eyes, a pineal opening (or 'third eye'), a single central nostril, and sensory fields, which well-preserved fossils show to have been connected to

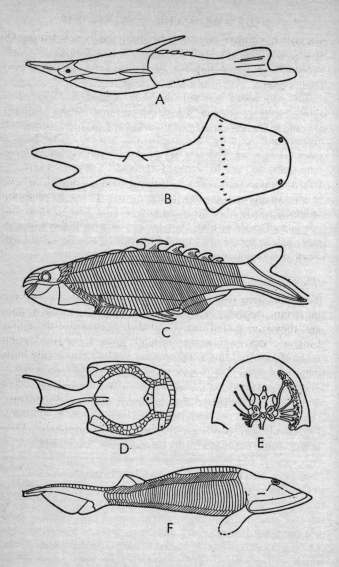

THE RISE OF THE VERTEBRATES

the rather primitive vertebrate brain by thick nerve trunks. On the underside of the head was a mouth and a series of ten gill openings. On the posterior border of the head shield were two pectoral flaps.

The cephalaspids seem to have been bottom dwellers, grubbing in the mud of streams and lakes, but other agnatha, the anaspids, such as *Birkenia* (Fig. 22C), lacked this heavy armour and were probably active swimmers. These and the ostracoderms were clearly adapted to a variety of different environments, yet in spite of their diversity, they are not found after Devonian times. Indeed, the only agnatha known at all from post-Devonian times are the living hag-fish and lamprey, which have probably survived because of their very specialized mode of life. They feed by attaching themselves to other fish by means of a funnel-like sucking mouth, armed with a rasping tongue, by which they feed on their victims.

The placoderms

The *Placodermi* (Fig. 24) differ from the agnatha in two very important respects – they have jaws (though primitive ones) and they have paired fins. Until their appearance the opportunities of development were severely limited, but the development of jaws and limbs represents a landmark, not only in the history of fish, but also in vertebrate evolution as a whole. Without jaws and paired fins the vertebrates could never have completely abandoned the mud-grubbing life of the ostracoderms, could never have taken advantage of new sources of food, could never have become the supremely successful aquatic group, could never later have left the water for the land.

FIG. 22 Palaeozoic jawless fish (Agnatha).
A *Pteraspis*, Devonian, length about 3 inches; B *Thelodus*, Silurian, length 4–8 inches; C *Birkenia*, Silurian, length 4 inches; D *Drepanaspis*, Devonian, length about 12 inches; E diagrammatic dissection of the head of a Devonian cephalaspid ostracoderm, showing brain, nerves, and sense organs; F *Hemicyclaspis*, Silurian, length 8 inches. (*A after White, B after Traquair, C after Stetson, D after Romer, E–F after Stensio*)

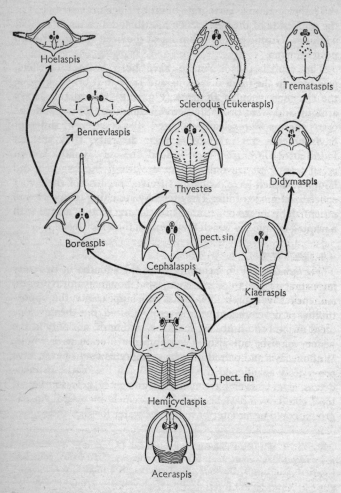

FIG. 23 Variation and development of head shields in cephalaspid ostracoderms. (*After Gregory*)

THE RISE OF THE VERTEBRATES

The placoderms' jaws were derived from the gill arches, the cartilaginous or bony structure which supported the gills. These jaws were primitive in comparison with those of later forms, in that the hyomandibular (the upper portion of the first gill arch) was not incorporated in the suspension of the jaws, as it is in all later fishes.

'It is immediately apparent that the transformation of a gill arch into jaws was a natural evolutionary development, perhaps the simplest possible solution to the basic problem of developing a pair of vertebrate jaws.' So writes a very distinguished palaeontologist – and this is true, but it is true only in retrospect. It was a 'natural' development only because we now know that it took place. Seen in prospect rather than retrospect, it is a wonderful thing that over long periods of time structures may undergo such revolutionary modification to perform completely new functions. But this, as the same palaeontologist himself observes, is the way of evolution. We must later consider its significance.

The placoderms appear in the Upper Silurian, but their ancestors are not known. They probably originated, however, from primitive agnatha. They include a number of very different marine and fresh-water forms varying from small spiny creatures a few inches in length to the gigantic arthrodires which were undoubtedly the ruling group of their day. The fresh-water acanthodians or 'spiny sharks' are a very primitive group (Fig. 24 D).

The arthrodires were much more formidable, and although some were quite small they included one gigantic genus *Dinichthys* which reached a length of more than thirty feet (Fig. 24 B). They were characterized by a massive armoured head, in which the skull articulated on a pair of ball-and-socket joints against the thick shoulder armour. These jointed-neck carnivorous monsters thus seem to have opened their gaping mouths by moving the upper rather than the lower jaws, though the latter may have moved to some extent. The jaws were armed by sharp shearing bony projections of the skull. They were active predators and seem to have lacked

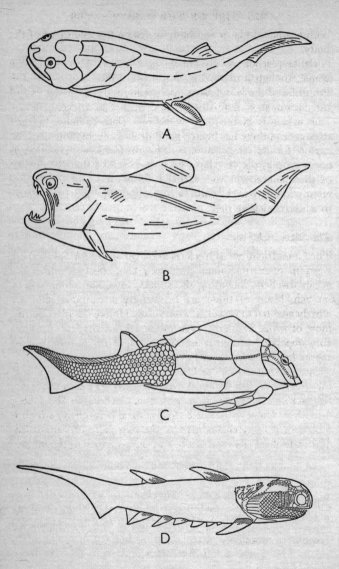

THE RISE OF THE VERTEBRATES

both armoured plates and bony scales on the main part of the body.

Other placoderms are shown on Fig. 24. The group as a whole, though it marked such a major advance in the overall evolution of the vertebrates, was greatly reduced at the end of the Devonian and finally became extinct at the end of the Palaeozoic. It is the only vertebrate class to have become extinct, and there has been a good deal of speculation as to the reason for its disappearance. The problem of extinction is never an easy one (p. 189), but it may well be that the decline of these more primitive forms is connected with increasing competition from the more advanced sharks and bony fishes to which they gave rise.

The Chondrichthyes

The Chondrichthyes (Greek *chondros* cartilage: *ichthys* fish) are a group of cartilaginous, predatory fish, most of which are ocean dwellers, including the sharks, rays, and chimaeras or rat-fish. Many of them are beautifully streamlined and are clearly adapted to powerful swimming. Unlike the placoderms, most of which had large eyes and small nostrils (and presumably depended primarily upon their sight for hunting), most sharks have small eyes and well-developed nostrils (Fig. 25).

Some living sharks reach a length of fifty feet, and one ray, the great devil fish (*Manta*), may measure as much as twenty feet across. In comparison with the placoderms the Chondrichthyes show an increase in structural complexity which is represented by such features as the two pairs of paired fins, more advanced jaws and teeth, and the structure of the ear and

---

FIG. 24 Placoderms.

A Reconstruction of *Coccosteus*, a Devonian arthrodire, length 18 inches; B reconstruction of *Dinichthys*, total length about 30 feet (*after Parker*); C *Pterichthyodes*, a Devonian antiarch, length 6 inches (*after Traquair*); D *Climatius*, a Devonian acanthodian, length about 3 inches (*after Watson*).

THE EVOLUTION OF LIFE

the reproductive organs. The eggs are usually fertilized internally, and male sharks bear clasping devices on the pectoral fins. Sharks do not have lungs or air bladders and they lack true bone, their skeletons being cartilaginous. The group is well represented in the geological record, however, because of their

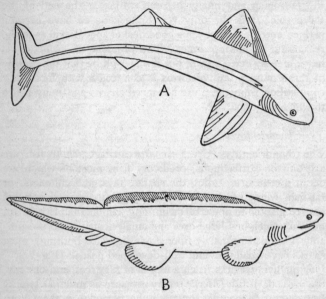

FIG. 25 Palaeozoic sharks.
A *Cladoselache*, a Devonian shark-like fish, 3 feet in length (*after Marchand*); B *Pleurachanthus*, a late Palaeozoic shark, about 2½ feet in length (*after Tenison*).

resistant teeth and spines. The body is often covered by minute dermal denticles or scales.

The sharks first appeared in the Devonian and underwent a considerable expansion in the Upper Palaeozoic. Their earliest members were fresh-water forms, but almost all later types were marine. At the end of the Permian, however, very many forms became extinct, and the succeeding members of the

THE RISE OF THE VERTEBRATES

group were less numerous. Living members are, however, a successful group, in spite of their small number of species.

Some of the earliest known sharks are remarkably like their living representatives. One of the best known is *Cladoselache* (Fig. 25 A), a creature of about three feet in length, from the Upper Devonian of Ohio. It is found in marine black shales, where even some of the soft parts of the body have been preserved. The jaw suspension of *Cladoselache* is primitive, although it is still found among a few living forms, such as the Port Jackson Shark of Australia (*Heterodontus*). It seems probable that *Cladoselache* does not differ greatly from the types of fish which gave rise to the various other members of the group, although it is itself too late in time to be an ancestral form. Most of the later members (sharks, skates, rays, and so on) live in the seas, though one extinct group (the pleuracanths) were fresh-water dwellers. Typical sharks are supremely well adapted for a predatory life in the open waters while more recent skates and rays, on the other hand, represent an extreme adaptation to bottom-dwelling. In them the pectoral fins are greatly enlarged to form wing-like structures and the teeth are highly modified for crushing shell-fish.

The Osteichthyes

The Osteichthyes are the familiar bony fish of the seas and streams and pools. They differ from the Chrondrichthyes in their bony skeletons, their scales or plates, the possession of an air bladder (formed by modification of the more primitive lungs), and the complexity of the brain. Many extinct forms were heavily armoured. The body is very variable in form (Fig. 26) and members inhabit a great diversity of environments. They outnumber all other living fish by about twenty to one, and in fact outnumber all other vertebrates combined. Yet it is not in numbers alone that the bony fish are so successful, for they inhabit an extraordinarily wide range of environments – from depths of more than 15,000 feet to heights of 14,000 feet in the Andes, from the poles to the equator, from hot spring waters to freezing ponds and lakes and even caves. Most fishes are restricted to either fresh or marine water, but some (such as sticklebacks) can survive both and other salt-water forms migrate to fresh water for spawning (e.g. salmon) or vice versa (e.g. the common European eel).

The oldest fossil bony fish are found in the Middle Devonian and most of the early representatives of the group are freshwater forms. From the outset there appear to have been two more or less distinct types of bony fish. The first of these is the actinopterygians or ray-finned fish, and this includes both the most primitive members and also almost all living members of the group, in which the fins are composed of a web of skin supported by many slender horny rays. Most of the members lay enormous numbers of eggs, and this may not be the least important factor in the success of the group.

Throughout the Palaeozoic the ray-finned fishes were a rather insignificant fresh-water group. The Mesozoic saw a rapid expansion of the group and their wholesale invasion of the seas. It was from these creatures that the hosts of living bony fish (the telecosts) of the seas and inland waters were derived. A diagrammatic sketch of their development is given in Fig. 26.

The most important trends in their development were the replacement of the heavy rhombic scales by thinner, rounded ones; the complete ossification of the skeleton together with changes in jaw structure; the replacement of the primitive heterocercal tail structure (with the vertebral column flexed up into the upper lobe as still found in the primitive sturgeon) by a homocercal condition (with more or less symmetrical lobes and the vertebral column ending near the middle of the base), and the replacement of the primitive lungs by an air bladder. These were the trends that ultimately led to the shoals of millions of herring, the electric eels and sea horses, the flying fish of warm seas, the clown-like globe fish, the delectable trout of mountain streams, the exotic creatures of the aquarium, and the grotesque, light-producing fishes of the ocean depths.

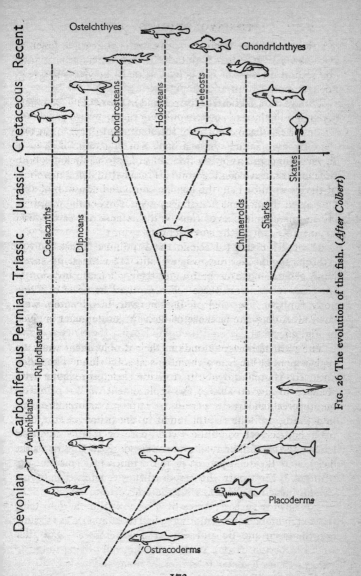

FIG. 26 The evolution of the fish. (*After Colbert*)

The Choanichthyes

The second division of the bony fish is the Choanichthyes, a much smaller group containing air-breathing fish, most of which have internal nostrils (or choanae). They are not the only fish to develop lungs (a couple of living but primitive genera of ray-finned fish have them), but the significant thing about the Choanichthyes is that these nostrils opening into the mouth are not found in any other fish but are essentially similar to those of the terrestrial vertebrates. Their living representatives are so few as to be almost insignificant in comparison with the ray-finned fishes, but their Devonian forebears occupy a unique place in the evolution of the terrestrial vertebrates. Within the Choanichthyes we can recognize two distinct groups, the lungfish (or Dipnoi) and the lobe-finned fish (or Crossopterygii). The lungfish are still represented by three living genera – one each from Australia, Africa, and South America – of which the Australian genus, *Epiceratodus*, is the most primitive. This creature lives in rivers in Queensland, an environment similar to those of both the other genera in the marked effects of seasonal drought. During these dry seasons the lungfish, although it cannot live out of water, is able to survive by breathing air by means of its single lung. The more advanced African and South American lungfish, which have two lungs, are even able to live in mud burrows out of water for quite considerable periods of time. Furthermore the Australian lungfish uses its paired fins to crawl along the bottom.

This introduces us to the second difference between these fish and the ray-finned types, for the slender web-like fin frame of the latter is quite different from the much more massive fin supports of the Choanichthyes consisting of an internal bony skeleton which articulates with the limb girdle.

Now how do such creatures fit into the evolutionary history of the vertebrates? It is tempting to hail them at once as missing links between aquatic and terrestrial vertebrates – but this proves to be too hasty a verdict. For not only living lungfish but even their Devonian forebears are altogether too specialized

THE RISE OF THE VERTEBRATES

to fill such a role – they all show reduction in the amount of bone in the skeleton, the bones of the skull are quite unlike those of other vertebrates, the teeth are highly modified for crushing food, and so on. But all in all, the lungfish, although they were not the direct ancestors of the land-living vertebrates, have affinities which suggest that they were probably quite closely related to them.

Lobe-finned fishes

We saw that the other group of choanate fishes are the lobe-finned crossopterygians. The group first appeared in the Devonian and was believed until recently to have become extinct in the Mesozoic 70 million years ago. In 1938, however, a living coelacanth was caught off the coast of South Africa in the Mozambique channel, and others have subsequently been captured.

Now, although the coelacanths are certainly crossopterygian fishes, they are not typical of the group, for the earliest members and virtually all the Palaeozoic choanates were freshwater rather than marine creatures. They were abundant in Devonian times, although they did not survive the Palaeozoic. These early forms (such as *Osteolepis* and the rather more advanced *Eusthenopteron*) were powerfully built, bony-scaled, streamlined carnivores, and they differed in a number of important respects from the lungfish. Their skeletons, for example, tend to contain more bone. The skull and jaws are completely ossified and the pattern of the bones of the skull is quite unlike that of the lungfish; their diet seems to have been more general and the teeth lack the crushing specializations of the lungfish. Their fins too have a different structure.

But the lobe-fins have other affinities than those with the lungfish. Suppose we compare one of them (*Eusthenopteron*, Fig. 27) with the early amphibians. The general skeletal plan is closely similar: the pattern of bones in the skull is essentially similar; the nostrils are similar; both the general form and the internal (labyrinthic folded) structure of the teeth are similar; the strong single central bony axis of the lobe-fins, fringed with

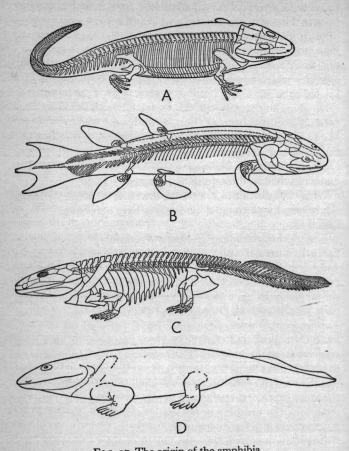

FIG. 27 The origin of the amphibia.
A *Diplovertebron*, a small Carboniferous labyrinthodont amphibian; B *Eusthenopteron*, a Devonian lobe-fin crossopterygian fish (length about 2 feet) showing the general similarity in structure; C–D *Ichthyostega*, a Devonian amphibian, about 2½ feet long: C skeleton, D reconstruction. (*A–B after Gregory, C–D after Jarvik*)

a. Stumps of the giant lycopod trees *Lepidodendron*, preserved as natural casts in sandstone. Fossil Grove, Victoria Park, Glasgow

b. (left) Track of dinosaur in Lower Cretaceous sandstone of Texas, with small boy sitting in 'bath'. *c.* (right) Skin of Cretaceous duck-billed dinosaur, preserved by desiccation and later buried in drifting desert sand

a. The flesh of this ichthyosaur, a marine reptile of Jurassic age, has been preserved as a carbonaceous film

b. A bony fish, *Portheus*, of the Cretaceous period. Length 12 ft.

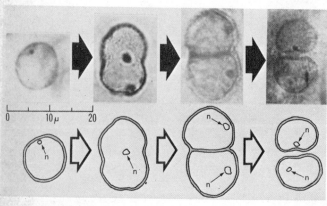

c. Glenobotrydion aenigmatis Schopf, from the Precambrian Bitter Springs Formation of Australia. The specimens show apparent stages of cell division, and the dark bodies (n) are believed to represent cell nuclei.

a. Nest of six dinosaur eggs from Cretaceous sandstone of Mongolia

b. A reconstruction of the 'Burgess Shale' Middle Cambrian sea, showing two colonies of branching sponges, several kinds of small crustacean-like arthropods, trilobites, sea cucumbers, and jellyfish

c. Spriggina floundersi Glaessner, which is a worm-like form; a specimen of the Edicara fauna from South Australia.

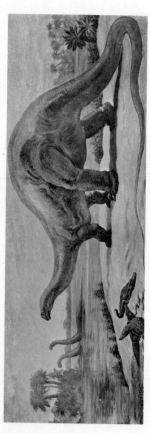

a. Apatosaurus, a Jurassic sauropod dinosaur, 67 ft long, with an estimated weight of 30 tons

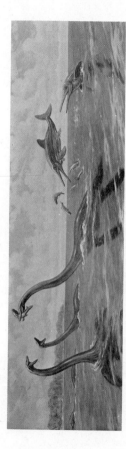

b. Jurassic marine reptiles. *Plesiosaurus* and the dolphin-like *Ichthyosaurus*. Some plesiosaurs reached a length of 50 ft and some ichthyosaurs 30 ft

A Cretaceous scene, showing a variety of dinosaurs. Left to right: *Struthiomimus* (an 'ostrich' dinosaur); *Triceratops* (a 20-ft-long ceratopsian); *Tyrannosaurus* (a 47-ft-long carnivorous theropod); *Ankylosaurus* (an armoured dinosaur; and *Anatosaurus* (a semi-aquatic duck-billed dinosaur). Also shown is *Pteranodon* (a flying reptile, with a wing span of 27 ft). Angiosperm plants include (left to right) palms, magnolia, oak, dogwood, laurel, and sassafras

The primitive hoofed mammal *Ectoconus*, a sheep-sized condylarth, of Eocene age

A Miocene prairie in North America. Left to right: *Diceratherium*, a small rhinoceros; *Promerycochoerus* and *Merychochoerus*, amphibious oreodonts; *Moropus*, a horse-like, clawed chalicothere; *Machairodus*, a large sabre-tooth cat; *Dinohyus*, a giant pig, 6 ft high; herd of *Merychippus*, a pony-sized horse; *Syndyoceras*, a double-horned, deer-like creature; *Procamelus*, an ancestral camel; *Merychyus*, an oreodont; *Alticamelus*, a giraffe-necked camel; *Gomphotherium*, a mastodon

At the dawn of the Ice Age the tall grasses were replaced by shorter, hardier vegetation. Left to right: *Mastodon*, about 10 ft in length; *Castoroides*, a large beaver; *Teratornis*, a condor-like bird with a 12-ft wing span; woolly mammoth; the surviving musk ox; *Bison crassiocornis*, whose horns had a span of 6 ft; *Canis dirus*, a wolf; *Smilodon*, the greatest of the sabre-tooth cats, larger than a modern tiger, with six-inch fangs; *Equus*, a 'modern' horse; *Boreostracon*, a giant armadillo-like glyptodont; *Megatherium*, a 20-ft high ground sloth

dermal rays, is strikingly similar to the limb of primitive amphibia. Even the environment in which these fish lived is strongly suggestive – a fresh-water environment that seems to have been characterized by seasonal drought. Here surely is a major clue to the ancestry of the land-living vertebrates. The resemblances are overwhelmingly strong. Certainly these fish *could* have given rise to the amphibia. But the question is 'Did they?' Can we find any transitional forms between amphibia and fish that raise a strong suspicion to a demonstratable fact?

LIFE ON LAND AND WATER

We have already spoken a good deal about amphibia. They represent the simplest of the group of terrestrial vertebrates, known as the tetrapods, which also includes reptiles, birds, and mammals. The amphibia are represented by the living salamanders and newts, toads and frogs, caecilians, and a variety of fossil forms. They were the first great group of vertebrates to adapt themselves to life on the land, although, as their name (Greek *amphi*, dual: *bios*, life) implies, they also live partly in water. In structure they stand between the fish, from which they arose, and the reptiles, of which they were the ancestors. Now the change from an aquatic to a terrestrial life is clearly a very fundamental one, which must involve major changes in body structure to meet the new conditions. We therefore find very marked differences between fish and amphibia, and almost all of them are associated with this new way of life. Thus amphibians develop two pairs of limbs (in place of fins), and lungs with nostrils that open into the mouth cavity and have valves which exclude water (in contrast to the gills of fish, although gills are present in the larvae of amphibia). The skeleton too needs to be stronger to support the weight of the body in air; the moist, glandular skin must permit exposure to the air, and changes are also involved in the circulatory and nervous systems. These profound changes marked the first vertebrate conquest of land, but the liberation of the amphibia from the water remained incomplete in one respect – a respect of the utmost importance: most of them need to return to the water to reproduce and almost all of them are restricted to moist environments.

Although the common frog is an atypical amphibian in its body form and locomotion, its life history provides an illustration of both the advance and the limitations of amphibian life. Most frogs live in damp environments of temperate and tropical areas (though there are exceptions), their active lives (reproduction, growth, and feeding) being confined to the warmer seasons, and, in the colder areas, the winters being spent in hibernation. The eggs lack a hard protective covering and are laid in water, the male fertilizing them as they are shed by the female. These eggs contain a relatively small food supply and rapidly hatch into larvae or tadpoles. These familiar long-tailed creatures have three pairs of external gills on their egg-like heads, but these are later replaced by internal gills, with gill slits. This is the beginning of such a radical metamorphosis that the whole character of the animal changes. Hind legs develop, later to be followed by forelegs; lungs develop and the animal begins to gulp in air; the gills and tails are resorbed, the intestinal tract is modified, and gradually the whole animal changes from fish to frog. Even the irate parent, tormented by his children's regular springtime collection of tadpoles, cannot ignore the beauty and the wonder of this transformation.

Something of the same transformation is revealed by the oldest fossil amphibia, the ichthyostegids from the Upper Devonian of east Greenland, for these remarkable fossils represent creatures which in a number of ways are intermediate between crossopterygians and the typical early amphibia (Fig. 27).

More than 200 specimens are known and two genera appear to be represented. They were squat sprawling creatures about three feet in length. Their heavy flattened skulls had roofing bones which bore many similarities to those of such crossopterygian fish as *Eusthenopteron*, but there were also differences and these were more suggestive of the skull of later tetrapods – some of the bones were lost or reduced, the eyes were more

THE RISE OF THE VERTEBRATES

centrally placed along the length of the skull, and there were slight changes in the nostrils. But the mixture of fish and amphibian characters was not confined to the skull, for, although the animal was typically amphibian in its strong pelvic and pectoral girdles, with well-developed limbs and five-toed feet, it retained the primitive vertebrae and the fish tails of the crossopterygians. Even the rays which support the tail fin are characteristic of fish and quite unlike the unsupported median fins of some later amphibia.

There were other less conspicuous structures too in which ichthyostegids were a mixture of both fish and amphibia. The lateral line tubes (a system of sense organs enclosed in canals, which detect pressure changes) which perforate the skull are characteristic of fish, as are other features of the skull; but the unjointed skull, free from the pectoral girdle and its relative proportions and the 'double-headed' ribs are characteristic of amphibia.

We must be careful, however, not to carry the conclusion too far, for the ichthyostegids, for all their mixture of characters, were probably not the direct ancestors of all the later amphibia. In some of their characters they are already more specialized than later amphibia, which developed similar characters (those associated with the articulation of the skull and the vertebral column, and the position of the articulation of the jaws) only at a later period of their history.

If any animals ever deserved to be regarded as missing links, it is these ancient ichthyostegid creatures. Amphibians adapted to life on land, to be sure, yet retaining such an imprint of their forebears that there can be no doubt of the path by which they evolved. Another creature (*Elpistostege*) intermediate between fish and amphibia is known only from a skull found in Quebec. But from our view-point the ichthyostegids are more than a mere group of interesting fossils. They represent the first known conquest of the land by vertebrates; indeed, although they themselves are not the direct ancestors of all later tetrapods, they are almost certainly broadly similar to the creatures from which all the subsequent land vertebrates arose. It is to some such group as them that we may trace our most remote forebears. Yet their life on the land is a 'conquest' only in retrospect: in its origin it was almost certainly more an infiltration or even a strategic withdrawal than a conquest, for, ironically enough, there seems a strong possibility that these creatures were driven ashore by the necessity of finding new food supplies and perhaps new pools or streams of water in the conditions of seasonal drought under which they appear to have lived.

EARLY AMPHIBIA

We have already seen something of the general pattern of amphibian structure and its relationship to the peculiar problems of life on the land. We must now turn to the early amphibia and trace their development. They include a startling variety of forms, but most of the ancient amphibia belong to a broad group known as the labyrinthodonts or the stegocephalia, the first name indicating the labyrinthic wall structure of the hollow conical teeth (which they share with the crossopterygian fish), the second describing the heavy roof of the massive skull. The group became widespread during the Lower Carboniferous and survived until Triassic times, a period of about 100 million years, and it included numerous forms with a world-wide distribution. Although some of its members were well adapted to land life, others were not. One important group, the embolomeres, were rather fish-like in appearance and about fifteen feet in length, with weak limbs which suggest that they spent most of their lives in the water. This primitive group of fish-eaters was common and widespread in Coal Measure swamps (Fig. 28).

It was probably from some such forms as these that the best-known labyrinthodonts evolved – the giant land carnivores known as rhachitomes, such as *Eryops* (Fig. 29). These were squat creatures, with heavy, flat, triangular skulls, large mouths, and sharp, strong teeth in both the jaws and the palate. The ornamented surface of the skull suggests that it lay at or

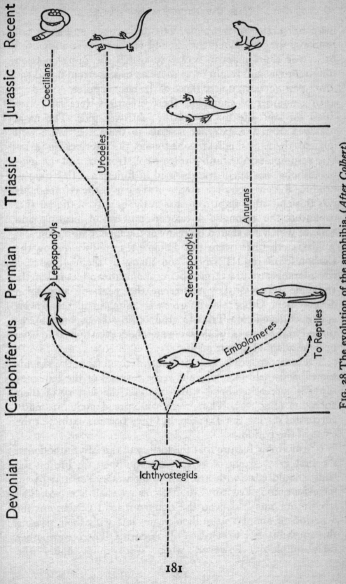

FIG. 28 The evolution of the amphibia. (*After Colbert*)

THE EVOLUTION OF LIFE

near the surface of the body. The bulky body and short tail probably dragged along the ground for they were supported by rather short massive limbs, in which the upper portion (the humerus and femur) was more or less horizontal and the lower part vertical, unlike those of the later reptiles.

In a number of characters these creatures were far better fitted for land life than were the ichthyostegids. The nasal passages from the external nostrils to the throat were well defined, the fishy medial fins had gone, the interlocking vertebral column and the limb girdles were stronger, and the hyomandibular was again transformed in function – this time to produce the characteristic stapes of the ear of the tetrapods. It was this and other improvements in the quality of the ear that produced the strongly developed powers of hearing and balance which are shared by many of the higher vertebrates.

These creatures were the dominant amphibia during the Carboniferous and Permian and many of them grew to a considerable size, the largest being about fifteen feet in length. In some respects they represent the peak of amphibian development. Other related forms were much smaller, however, and one degenerate Triassic group (the stereospondyls) returned to the waters, while some members even appear to have become adapted to marine life.

The stereospondyls were probably derived from the rhachitomes, and their skeletons showed a reversal of the tendency towards increasing bone which marked the history of their ancestors (Fig. 28). They were widespread and apparently successful during the Triassic, but they became extinct at the close of the period.

It is an ironic feature that the history of the labyrinthodonts, the first great group of vertebrates equipped for life on the land, should close with a return to the water. Yet it is by no means a unique event in evolution, as we shall later see. The labyrinthodonts had been the dominant land vertebrates for a period of about 100 million years, but with their passing the amphibia fell to a place of obscurity. There were other early amphibia, however, which were quite unlike the

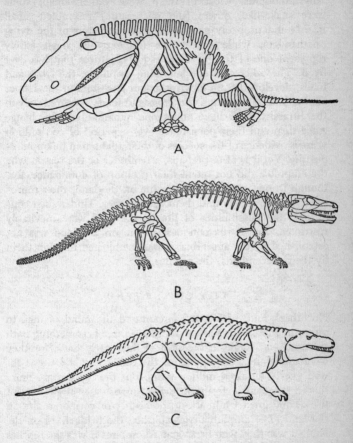

FIG. 29 A *Eryops*, a Permian labyrinthodont amphibian, about 5 feet in length (*after Colbert*); B–C *Seymouria*, a Permian 'progressive amphibian or primitive reptile' about 2½ feet in length: B skeleton (*after T. E. White*), C restoration (*after Gregory*).

labyrinthodonts. Most of them were rather small, some were snake-like, others (the microsaurs) resembled small lizards and probably gave rise to the ancestors of the living salamanders, while one genus (*Diplocaulus*) must surely represent one of the most grotesque creatures that has ever lived. The other common surviving amphibia, the frogs and toads, developed from labyrinthodont ancestors in the Upper Carboniferous, although true frogs and toads appeared only in the Jurassic. They have undergone remarkably little change since then and their persistence over a period of 200 million years is evidence of the success of their adaptation to amphibious life. Yet it is also, perhaps, a reminder of the reason why the amphibia did not retain their position of dominance, for, though in one sense adapted to life on the land, their reproductive cycle still bound them to the water. Under such conditions, the possibilities of life on the land were inevitably restricted. But if the conquest of life on the land was not established by the amphibia, it was established through them by their descendants, the reptiles.

LIFE ON THE LAND

No other group of fossils has captured the mind of man to quite the same extent as the reptiles – there is something both splendid and tragic in the saga of their early days, for they were undisputed champions of the earth, air, and water for well over a hundred million years. But the reptiles are more than a spectacular group of ancient monsters: they represent the second great landmark in the history of vertebrate life on the land. The amphibians established the bridgehead on the land; it was they who first went ashore, but it was the reptiles that stayed ashore. The amphibia evolved from the water, their members (except for the specialized caecilians) were born in the water, most of them lived near its edges in order to retain a moist skin, they returned to the water to reproduce: they were the heirs of the land, but they never ceased to be the children of the water. The amphibian achievement was life on

THE RISE OF THE VERTEBRATES

the land and the water: the reptilian achievement was life on the land.

The essence of this change lay in the manner of reptilian reproduction. We have already seen that the amphibian developed in the water, from which it derived oxygen, food, and protection against both desiccation and injury. The development of the reptile (or amniote) egg provided the same

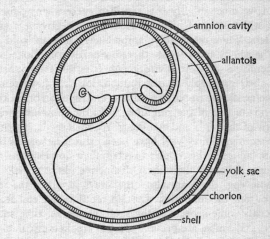

FIG. 30 Generalized diagram of the embryo of higher vertebrates. (*After A. S. Romer*)

benefits on the land or within the mother's body, and the development of copulatory organs allowed internal fertilization.

The developing embryo, which is enclosed in a sac of liquid (the amnion), obtains nourishment from a large yolk and is attached by the end of the gut to the allantois, which functions as both an embryonic bladder and a lung against the porous shell.

The other reptilian characteristic was the possession of a dry, cornified 'skin', often covered with overlapping scales. It was this protective covering that provided a means of solving the

THE EVOLUTION OF LIFE

second great problem of land life, the loss of moisture by evaporation through the skin.

With these changes went others. The limbs are suited for more rapid locomotion than those of amphibia and are variously modified for grasping, climbing, flying, and swimming in some forms, but in others (snakes) they are lost altogether. The heart is so constructed that the circulatory system becomes more efficient.

But we must qualify this reptilian success story, for there were still features which reduced their effectiveness as land dwellers. They retain their amphibian forebears' cold blood and their body activities are therefore partly conditioned by the temperatures of their environment. It is this factor that controls their geographical distribution, for they have remained characteristically tropical creatures. The reptilian brain, though superior to that of the amphibia, lacks the complexity characteristic of the birds and mammals. We shall see later that these 'shortcomings' may well have been critical factors in the ultimate failure of the reptiles to retain their dominance.

Now, although their most distinctive characteristic is the amniote egg, there are also other ways in which the reptiles differ from the amphibia. Certain reptilian skeletal features, especially some aspects of the skull, vertebrae, ribs, shoulder girdle, and limbs, are peculiar, and so are their circulatory and muscular systems. All these differences are clearly defined in living forms, and by means of them it is comparatively easy to distinguish an amphibian from a reptile. In the case of some fossil tetrapods, however, there is not such clear distinction. We find, in fact, that there were creatures possessing a mixture of characters – some clearly amphibian, others no less clearly reptilian.

This is well illustrated by *Seymouria* (Fig. 29B), a squat creature about two feet in length, from the Lower Permian of Texas, which has given its name to a group of related forms (the seymouriamorphs). The structure of its skull, teeth, and some features of the vertebrae are typically amphibian. In other respects, such as other characters of the vertebrae and

THE RISE OF THE VERTEBRATES

jaw suspension, and the structure of the shoulder girdle and limbs, however, *Seymouria* is more like a reptile.

But what is *Seymouria*? Is it a reptile, or is it an amphibian? At present we do not know – some palaeontologists consider it to be one, some another. The clear answer could be given only if we knew something of the creature's reproductive habits. If it laid an amniote egg on land, it was a reptile: if it did not, it was an amphibian. The answer is as simple, and as difficult, as that.

Seymouria itself is both too recent geologically and too specialized to be the direct ancestor of the reptiles. But it is probably not greatly different from those ancestors, and it illustrates an important aspect of evolution, namely that the gradual emergence of what we regard as typical reptiles was a piecemeal process. There was no integrated overall evolution of the typical reptilian features – the typical characters emerged not all together, but at different times, one by one, some earlier, some later – it was only when the last appeared that the 'typical' pattern found expression. The intermediate creatures did not have characters which were half amphibian, half reptilian. They were instead a jumble, with some advanced characters which were fully reptile and with others which remained primitive and were fully amphibian. This, of course, should not surprise us, for it is precisely the pattern of evolution that we saw both in *Archaeopteryx* and in the ichthyostegids. It is the pattern of mosaic evolution, as de Beer calls it, and we shall later discuss its significance.

Chapter 8
THE END OF AN ERA

WE may now pause to glance back at the 300-odd million years of Palaeozoic history. We have traced the development of life from its obscure origin to the first appearance of fossils, through the hey-day of marine invertebrates – the dominance of trilobites and brachiopods, the rise of the cephalopods and corals. We have seen the lowly beginnings of the fish and their subsequent expansion, and followed the first tenuous foothold of life in the rivers and on the land – plants, eurypterids, scorpions and millipedes, snails, and clams. We have followed the emergence of the amphibia, the appearance of the reptiles, the spread of the lofty coal-forming forests, and the continuing richness of life in the ancient seas.

We have also described something of the background scene, for no drama can ever be quite separated from the stage on which it is played. The drama of life is no exception to this, for the earth has provided both the stage and the home of this procession of ancient life. But here the analogy breaks down, for the earth is more than the home: it is the source and womb of life, and also, incidentally, its grave. In this case it is not the plot and the actors who determine the background of the stage – it is the stage which 'determines' the drama – plot, actors, development, and all. This is not to imply some blind impersonal terrestrial force selecting and fashioning the forms of countless living things. But it is to imply that life exists within and is sustained by the fabric of the earth and its atmosphere – and apart from this planet we know of no living things. It comes therefore as no surprise to learn that some aspects of the development of life clearly reflect the changes in geographical and climatic conditions of ancient times.

We are all familiar with this interaction between living

THE END OF AN ERA

things and their physical environment. An unusually severe winter may bring death to countless animals and plants, migrating sand dunes will obliterate the vegetation of coastal areas and new plant species will repopulate the area, summer drought may bring mass death to wild animals, a hot summer may bring a plague of flies, a sudden freeze may bring catastrophic death to lake fish. Yet we must be careful not to oversimplify the relationship between life and its environment, for it is almost inconceivably intricate. The interaction of dozens of physical factors (temperature, pressure, light, density of water, salinity, winds, daily variations, seasonal changes, etc.) and the even more complex and subtle influences of other organisms – all these are varying components in the economy and the surroundings that mould and sustain living things. Almost any generalizations are therefore too general by half, and should be regarded with suspicion. With that warning – let us generalize!

We have already seen that in the geography as well as the life of the earth, the close of the Palaeozoic marked the end of an era. The renewal of widespread and intense earth movements in late Carboniferous and Permian times gave rise to new mountain chains in North America, Europe, and Asia and their rise was accompanied by widespread volcanic activity. In some areas, the earlier coal-swamp conditions persisted and around the lands the clear shallow seas still supported a rich fauna, although the seas were greatly restricted towards the close of the era. In other places diverse climatic conditions produced deserts and land-locked seas in which great thicknesses of salt and red sediments accumulated. Over what are now the lush areas of the southern continents, huge continental glaciers spread across the land.

Against this changing background many of the dominant Palaeozoic races dwindled. Permo-Triassic times have been called a 'crisis in earth history', and for many groups of living things they were just that. Some familiar groups became extinct – the once widespread trilobites, many of the earlier dominant brachiopods and bryozoans, the rugose corals, the blastoids,

THE EVOLUTION OF LIFE

the fusulinids, the lacy fenestellid bryozoans, as well as many of the characteristic cephalopods and crinoids.

Nor was the decline confined to the invertebrates. Amphibia underwent a drastic reduction, as did some of the fish, and such plants as the giant horsetails, the related creeping sphenophyllales, and many characteristic ferns became extinct. The scale-trees (lycopods) and cordaites of the coal swamps dwindled and later vanished, but from the latter arose the early conifers. Now not all these extinctions were simultaneous – many groups of plants became extinct before the end of the Permian – and it is temptingly easy to over-simplify and dramatize this 'catastrophic dying'. Yet having said this, it remains true that such mass extinction is almost unique in earth history (only that of the Upper Cretaceous is comparable in magnitude), and it seems altogether too difficult to believe that it was not a reflection of extreme Permian environments. The widespread glaciation in the Southern Hemisphere, with high emergent continents, late Palaeozoic mountain building in many areas, the spread of reefs, red beds, and salt deposits, and the great restriction of Late Permian seas – these seem to have formed a combination of features which produced a profound, though intensely complex and varied, series of effects upon Upper Palaeozoic life.

But one of the features of the history of life is the tendency for new groups to colonize newly vacated environments, to replace the lost races. It was this evolutionary surge of new forms that went hand in hand with the mass extinction of the late Palaeozoic, and with these new forms there dawned a new era in life's long history – the Mesozoic, the days of the reptiles.

Chapter 9
THE DOMINANCE OF THE REPTILES

THE LAND

WE have already seen some of the ways in which the reptile body differs from that of the amphibian, and in terms of mode of life and range of possible habitats the difference is equally profound. The amphibia were limited to swamps and lowland marshes; in these environments the reptiles replaced and excelled them in their adaptation. But the advent of the reptiles also marked the vertebrates' invasion of other environments, some totally new – forests and deserts, the seas and the air. It is this widespread adaptation and diversity over a period of one hundred and fifty million years that is the measure of the reptiles success. In the long history of life only two other groups, the mammals and the birds, have achieved anything approaching such a diversity of adaptation; and even their success is an indirect tribute to their reptilian ancestors.

Stem reptiles

The most primitive reptiles are the cotylosaurs (a name referring to their cup-shaped vertebrae) which probably developed from the ancestral seymouriamorphs and which first appear in the Upper Carboniferous. It was from this group of 'stem reptiles', as they are frequently called, that the later conquerors arose, although the group itself became extinct in the Triassic. By Lower Permian times the cotylosaurs had become diverse and widespread, and two more or less distinct types may be recognized.

The more primitive captorhinomorphs (a name describing their 'gulping' jaws) were generally rather small (about a foot long) sprawling carnivorous creatures, with flattened skulls and pointed teeth.

The diadectomorphs ('broad [-toothed] forms') were larger

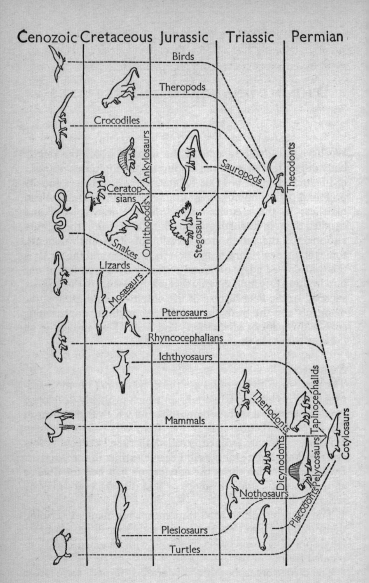

THE DOMINANCE OF THE REPTILES

(usually about six feet in length) with specialized skulls and teeth. Some were similar to the captorhinomorphs in general appearance, but others were massive creatures, standing four feet or more in height and reaching ten feet in length. These pareiasaurs ('helmet cheek reptiles', Fig. 32) were probably vegetarians and lived in the Permian swamps of Russia, Scotland, East and South Africa. Their weight, which must have been considerable, was supported by rather vertical limbs. The pareiasaurs did not give rise to the main group of reptiles, but it is worth pausing to reflect upon the importance of their vegetarian habits in the history of the vertebrates, for they represent another landmark in the route by which the higher vertebrates gained their supremacy.

One of the features of the development of new major groups of organisms is the speed with which they diverge to produce a variety of different forms adapted to widely differing environments. This 'adaptive radiation' is particularly well shown in the reptiles, for although the cotylosaurs were the stem reptiles, many other groups arose so rapidly from them that they were the contemporaries of the later cotylosaurs before the ancestral group became extinct.

One group which appeared at a very early stage in reptilian history was the mesosaurs – long-tailed, slender, aquatic creatures, about three feet in length. Their long snouts were armed with a series of needle-like teeth and they seem to have been fish-eaters in pools and streams. Their remains are known only from the Upper Carboniferous and Lower Permian of South Africa and South America, and they do not appear to have given rise to any descendants (Fig. 32A).

Other early reptiles

There were, however, other early reptilian groups (appearing in Upper Carboniferous times) which certainly did produce descendants who were destined to play a major role in vertebrate evolution. The pelycosaurs ('basin-shaped [pelvis] lizards')

FIG. 31 The evolution of the reptiles. (*Based upon various figures by Colbert*)

THE DOMINANCE OF THE REPTILES

are a group of bizarre reptiles whose remains have been found in large numbers in the red beds of the American south-west and, more rarely, elsewhere. They were the first of a great group of reptiles known as synapsids, which were widespread during Permian and Triassic times, and which are a group of particular interest because it was from them that the mammals arose.

The pelycosaurs were a rather varied group, but most of them were large sprawling creatures with long tails. Many exceeded ten feet in length. Their most striking feature was the development along the back of greatly elongated neural spines, which in life appear to have been covered by skin and to have formed a sail-like structure (Fig. 32). Although not present in all members of the group, this fin was present in a number of quite different forms.

Dimetrodon was a slender active carnivore, about eleven feet in length, with a large, deep skull and powerful differentiated teeth, which had a 'sail' almost three feet high. *Edaphosaurus* belongs to a second group of pelycosaurs, superficially similar, but different in structure and habits; it was apparently a rather heavy and awkward herbivore, and the small skull contains a series of more or less uniform blunt teeth adapted to a vegetarian diet. But it too had a 'sail' – one in fact with horizontal 'yards' of bone growing from the main 'masts'. The function of these curious structures remains a mystery. Some have suggested they were protective, others that they were non-functional, perhaps the result of hereditary maladjustment, others that they were sexual characters, and others that they assisted in the control of body temperature.

FIG. 32 A *Mesosaurus*, an aquatic Lower Permian mesosaur reptile about 3 feet in length (*after Colbert*); B *Rutiodon*, a thecodont, Triassic phytosaur, about 12 feet long (*after Colbert*); C *Oligokyphus*, a small (2 feet long) advanced cynodont mammal-like reptile, Lower Jurassic (*after M. Wilson*); D *Dimetrodon*, a Permian pelycosaur reptile, length 10 feet; E *Cynognathus*, a mammal-like reptile from the Triassic of South Africa, about 8 feet long (*after M. Wilson*); F *Pareiasaurus*, a primitive Permian reptile, about 10 feet in length (*after M. Wilson*).

THE EVOLUTION OF LIFE

It was from the pelycosaurs that the other great synapsid group developed – the large and varied therapsid (mammal-like) reptiles which were widespread in Permian and Triassic times. Some members are strikingly similar in structure to the mammals, which descended from them. The teeth, for example, are strongly differentiated in later genera, and their skull, skeleton, and posture had many mammalian characters, although there were differences too.

Not all the therapsids showed such specialization as this, however. The dicynodonts ('dog-toothed') were a common and widespread Permo-Triassic group of large vegetarians, in many of which the teeth were greatly reduced (Fig. 31); some had tusks and others beak-like structures, while at least one genus seems to have been aquatic. Another group, the Permian dinocephalians ('huge heads'), were massive and powerful creatures, some of them reaching thirteen feet in length (Fig. 31).

It is in the theriodonts ('beast-toothed') that the mammal resemblances of the therapsids are most strikingly shown. These creatures were a group of active carnivores, whose remains are best known from the Permo-Triassic Karroo Beds of South Africa. The group was a varied one, but *Cynognathus* of Triassic age is a typical member (Fig. 32).

Cynognathus was an active creature, about six or seven feet in length. The skull was large, narrow, and long, and the teeth specialized and strongly differentiated into incisors, canines, and 'post-canines' or cheek teeth. Many other characters found in mammals were already developed: the separation of the mouth and nasal passage by a hard secondary palate, and several other features of the skull and postcranial skeleton (the vertebrae, limbs, hip bones, and shoulder blades, for example). The posture was distinctive, for the legs were placed below the body (knee forward – elbows back) and so lifted it clear of the ground, while the feet were fully developed and well suited for active locomotion. The creature still possessed, however, a powerful reptilian tail.

The Upper Triassic ictidosaurs ('weasel-lizards') from

THE DOMINANCE OF THE REPTILES

England and South Africa were tiny creatures which were even more mammalian in character – they were in fact regarded as reptiles only on the basis of their jaw articulation, and it has now been shown that at least one genus, *Diarthrognathus*, possesses, as its name implies, both the reptilian and the mammalian types of jaw articulation.

From such creatures it is but a short step to the mammals, and the structure of *Cynognathus* is another example of the way in which the division between related vertebrate classes becomes indistinct as we trace their earliest fossil representatives. Indeed it is possible that some cynodonts were even more mammalian than their fossil skeletons reveal – some may well have been warm-blooded and have possessed a hairy covering. The same sort of mixture of characters is still seen in the most primitive group of living mammals, the monotremes (p. 230).

The dinosaurs

But the day of the mammals was not yet, for the Mesozoic era, although it saw their rise, did not see their dominance. These were the days of the reptiles, and of this race of champions none was more mighty than the dinosaurs. To trace their origins we must turn to the order of reptiles known as the thecodonts ('socket-toothed') – a group of great importance, for it also gave rise to crocodiles, pterodactyls, and birds. The thecodonts were a widespread and varied group that flourished during Triassic times. They tended to be rather small animals, many of them protected by bony scales, and they had characteristic deep and light but strong skulls, usually with hollow teeth set in deep sockets (hence their name). One group of Triassic thecodonts (the phytosaurs – 'plant lizards', it being once thought that they were herbivores) were aquatic crocodile-like carnivores, some of which reached twenty feet in length (Fig. 32B). They lived in streams and lakes, much as modern crocodiles do, but for all their apparent similarity they were not the ancestors of the crocodiles. Their resemblance is the result, not of a close genetic relationship, but of similar adaptation to the environment in which both (at different times) lived – a good example of convergence, or parallel evolution, which we shall later discuss (p. 248).

The main group of thecodonts were quite different. They were small, active terrestrial carnivores, which were often bipedal in habit, and this change in posture led to far-reaching modifications throughout the body (Fig. 32). The rear limbs became large and strong, the whole weight of the body being concentrated upon the hip bones; a long balancing tail developed, and the fore-limbs were reduced to some extent.

It was from these creatures that the dinosaurs arose. The name 'Dinosauria' was coined in 1842 by Sir Richard Owen to describe three genera of large Mesozoic terrestrial reptiles. The word has since become a part of everyday language, but it is now clear that it includes two more or less distinct groups, the Saurischia or reptile-pelvis types and the Ornithischia or bird-pelvis types.

The dinosaurs are perhaps the most spectacular and familiar of all fossils, and throughout the long Mesozoic era they dominated the life of the land – and then vanished without trace. Most (but not all) of them were very much larger than their thecodont ancestors; many of them underwent a secondary return to the four-footed posture of their more distant forebears; some were active savage carnivores, others ponderous browsing herbivores; some were heavily and fantastically armoured, others were delicate bird-like creatures; some were swamp dwellers, others spent their days in upland areas. But for all their variety, they shared one common feature – they were the rulers of the Mesozoic lands.

The oldest saurischian dinosaurs are found in the Upper Triassic and they include the direct descendants of the thecodonts, the 'theropods' or 'beast-feet', a group of bipedal carnivorous creatures. Some of these were small delicate animals, only five feet in length, and these forms remained very similar to their thecodont ancestors. Other later forms were larger, ostrich-like, and possibly herbivorous in habit, with beak-

THE DOMINANCE OF THE REPTILES

like jaws. Yet others, however, were very different and included the giant carnivores: *Allosaurus* (Fig. 33A) from the Jurassic reached a length of thirty-five feet, and *Tyrannosaurus* from the Cretaceous was the largest land carnivore that has ever lived – fifty feet long and twenty feet high (Plate 5).

These monsters had powerful clawed rear limbs, with a heavy tapering tail, reduced fore-limbs with tiny, almost useless 'hands', and an enormous skull which was strong but light, with huge jaws and dagger-like teeth. In *Tyrannosaurus rex* ('the king of the tyrant lizards') the skull was four feet long, and three feet deep, with sharp curved teeth up to six inches in length. Mute testimony to the predatory habits of *Allosaurus* is provided by the scratched and broken bones of *Brontosaurus*, a large herbivorous dinosaur. The spacing and form of the grooves correspond exactly to the teeth in the skull of *Allosaurus*; indeed, broken teeth of this creature have been found with *Brontosaurus* bones.

The other saurischian group was the sauropods – a race of giant herbivorous quadrupeds which flourished during the Jurassic and Cretaceous. These were the largest animals that have ever lived on the land; some of them were more than eighty feet in length and probably weighed fifty tons (Plate 4(*a*)). These creatures are represented by a number of genera found in various parts of the world. Most of them were broadly similar, and *Brontosaurus* and *Diplodocus* are typical. The bulky body was supported on four pillar-like limbs, of which the shorter fore-limbs are a reminder of their thecodont ancestry. The tail was extremely long and the long neck supported a tiny head. Several features of the skeleton suggest that these animals spent most of their time in lakes and rivers, browsing on the soft plants. The weight, for example, is concentrated in the lower parts of their great skeletons and this probably assisted their stability in the water, while the nostrils are placed with the eyes on top of the head, allowing the animal to breathe and see, even though almost completely submerged. This amphibious life probably afforded not only most of their food but also their best form of protection against their carnivorous

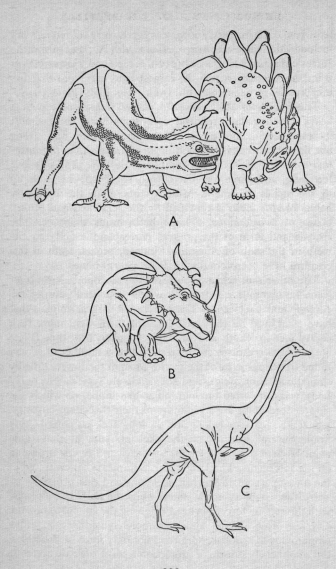

THE DOMINANCE OF THE REPTILES

enemies, and the buoyancy of their enormous bodies must have provided some relief for their overloaded legs.

The ornithischian dinosaurs are not found amongst the earliest dinosaur faunas, and are a rather more specialized group than the saurischians. They include four main groups, the relationships of which are shown in Fig. 34.

The ornithopods include the duck-billed dinosaurs and the forms from which they arose. They were all essentially bipedal animals, although their fore-limbs seem to have been sufficiently strongly developed to allow them also to walk on all four legs. One of the earlier members of this group is *Iguanodon*, a well-known dinosaur from the Cretaceous of Europe, which was the first dinosaur ever to be described (Fig. 34). *Iguanodon*, the thumbs-up dinosaur, was a bipedal herbivore, fifteen feet high and twenty-five feet long.

Some of the Cretaceous duck-bills are amongst the most grotesque of all fossils. They developed webbed feet, duck-like bills armed with as many as 2,000 teeth, and sometimes bizarre crested structures, and were well adapted to an amphibious life. Some of the crests, for example, contained long nasal passages which may have assisted in air storage during diving, while others functioned as 'snorkels'. Some related forms developed a grotesque thickening of the roof of the skull (Fig. 34).

The stegosaurs, mainly Jurassic, included a variety of armoured dinosaurs. *Stegosaurus* was a twenty- or thirty-foot-long quadruped, with a high arched back, along which two alternating rows of vertical triangular plates were embedded in the hide (Fig. 33). The heavy tail ended in four sturdy spikes. The tiny head contained an equally tiny brain (about the size of that

---

FIG. 33 Variety in dinosaurs.

A Two Jurassic dinosaurs, the carnivorous *Allosaurus*, a 35-feet-long theropod attacking an armoured *Stegosaurus*, a 20-feet-long ornithischian herbivore (*after Seyfarth*); B *Styracosaurus*, a multihorned ceratopsian from the Cretaceous, 20 feet long; C *Struthiomimus*, the ostrich-like, toothless, vegetarian theropod, with a beak-like jaw, length about 5 feet.

FIG. 34 Duck-billed dinosaurs, the ornithopods. These creatures were aquatic and web-footed, with variously shaped skulls as a result of the extension of the nasal bones. They were world-wide in Jurassic and Cretaceous times. The jaws of some contained 2,000 crushing teeth. A *Parasaurolophus*, Cretaceous, length about 30 feet (the crest contained a long air passage from the nostrils, allowing air storage when the animal was submerged); B *Trachodon*, Cretaceous, length 35 feet, a typical duck-bill (*after C. R. Knight*); C *Iguanodon*, Cretaceous, length 30 feet, a specialized ornithopod; D *Corythosaurus*, Cretaceous, length 30 feet.

THE DOMINANCE OF THE REPTILES

of a kitten) and the massive limbs and tail were controlled by local thickening of the spinal cord in the shoulders and hips.

The Cretaceous armoured dinosaurs or ankylosaurs were squat, tank-like creatures, such as the twenty-foot-long *Ankylosaurus*, with a broad armoured back and club-like tail. Their armour is reminiscent of the (later and quite unrelated) armadillos.

The horned dinosaurs or ceratopsians (Fig. 33 B) are confined to the Upper Cretaceous, but during their comparatively short span they displayed a remarkable diversity. They were all rather large herbivorous quadrupeds, with more or less massive bony head shields which were variously frilled, horned, and beaked. *Triceratops* (Plate 5), the last of their distinguished line, was a creature about twenty-five feet long, the great head shield accounting for no less than six feet of the total length. It was a herbivore, with a horny beak, but both the character of its head armour and the number of skulls found with healed fractures suggest that it was a savage opponent when attacked. One of the most interesting of all dinosaur discoveries is associated with this group of ceratopsians. In 1922 an expedition from the American Museum of Natural History discovered in Mongolia several exquisitely preserved clutches of eggs of the ancestral horned dinosaur *Protoceratops*. In two of the eggs unhatched embryos of the creature were preserved (Plate 3(*a*)).

Such were the dinosaurs, rulers of the earth for 150 million years. Their only real enemies were those of their own kind but for all their supremacy they were not to survive. They spread across the continents, the largest creatures ever to walk, yet the close of the Mesozoic was to mark their passing, and their race perished, leaving no survivors and no descendants.

Other reptiles

Although the dinosaurs were the most conspicuous they were by no means the only land reptiles of the Mesozoic, for three surviving groups were represented amongst the reptiles of that era (Fig. 31). The crocodiles were cousins to the dinosaurs, for they too arose from the thecodonts. The earliest member of the group was a small, heavily armoured, lizard-like creature, about three feet long (*Protosuchus*) from the Triassic. During the Jurassic and Cretaceous the crocodiles took over the environments formerly occupied by the phytosaurs and became both widespread and varied – highly adapted to the predatory life in streams and pools which still characterizes their descendants. The largest of them (*Deinosuchus* and *Rhamphosuchus*) reached a length of fifty feet, and some forms (the thallatosuchians or geosaurs, p. 207) even became adapted to life in the seas.

Although individual genera show a number of changes, the terrestrial crocodiles as a group have undergone little substantial alteration since the Mesozoic. They are still a successful, if limited, group, a living remnant and reminder of the days of the dinosaurs.

The rhynchocephalians ('snout-heads') are represented by the small living lizard-like reptile *Sphenodon* or tuatara from New Zealand. This creature is a true living fossil, the lone survivor of a great group of reptiles which first appeared in the Triassic and which has been a persistent, though relatively unimportant, group since then. It included the rhynchosaurs, of the Middle and Upper Triassic; they were large, heavily built animals, beaked and virtually toothless. It has been variously suggested that they ate molluscs on the sea-shore or dug up roots and tubers.

To most people reptiles mean snakes and (usually as an afterthought) lizards, and these two groups are certainly the dominant modern reptiles. The living iguanids, monitors, and chameleons are the products of an ancestral line that is first known in the Jurassic, and has persisted with some success ever since. During the Upper Cretaceous one group (the mosasaurs, p. 201) invaded the seas and included the greatest of all the sea-going reptiles.

The snakes are highly modified reptiles, in which virtually all traces of the limbs are lost. The delicate construction of the skeleton makes them rare as fossils, but they seem to be the

THE DOMINANCE OF THE REPTILES

youngest reptilian group, first appearing in the Cretaceous as an offshoot of the monitor lizards. The early forms were heavy (boa-like) types and fossil skulls suggest that the development of poison fangs was a much later (Miocene) development.

THE SEAS

One of the ironic things about the reptiles is that, having become the first group to establish themselves fully on the land, some of their number returned to the seas. In this respect they were not alone. We saw the same return to the water in the earlier amphibia, we shall see it later in the mammals and birds. It seems, in fact, one of the features of evolution that once any group becomes dominant in a particular habitat, it spreads into other environments, even those that its remote ancestors forsook.

The body changes involved in this secondary return are often every bit as profound as those involved in the earlier abandonment of a medium, but the end product usually bears little resemblance to its ancestor. The reptiles that returned to the seas did so, not as reconstituted fish, but as modified reptiles. For there is a sense in which the process of evolution is irreversible; once a structure has been replaced in a group the statistical probability that random genetic recombination and selection will ever give rise to anything fundamentally similar is infinitesimally small. This does not mean, of course, that there may not be a superficial resemblance between different forms sharing a common mode of life. This, as we saw with phytosaurs and crocodiles, is what evolutionists imply by the word 'convergence'. Sharks, dolphins, and the extinct ichthyosaurs lived the same kind of life in the same environment and they are all similar in outward form, but there the resemblance ends. For all their apparent similarity ichthyosaurs and dolphins are reptiles and mammals in fishes' clothing.

The diversity shown by the sea-going reptiles was just as great as that of those that remained ashore (Fig. 31). There were six types of reptiles that inhabited those ancient seas. The most familiar of these are the turtles, a common group amongst living creatures, and one that has undergone very little change since it first appeared in the Triassic, 175 million years ago. The earliest turtles were probably fresh-water amphibious creatures, of which a small number became secondarily marine in Jurassic times. Most turtles are still terrestrial.

The turtles seem to have developed from some such reptile as *Eunotosaurus*, a small creature only a few inches long from the Permian of South Africa, whose ribs were greatly enlarged to provide a partial upper shell. The earliest true turtles included forms with non-retractable heads, limbs, and tails, one genus of which had teeth. The teeth were lost in later forms, which developed beak-like jaws and a heavy shell, formed by the development of dermal plates, some of which were fused to the ribs. This protective shell seems to be the basis of the turtles' survival and they survive as a limited but successful group. They were one of the earliest reptiles to return to the sea, and they are the only survivors of that ancient armada. In Cretaceous times some turtles grew to a length of twelve feet, their huge paddle-like limbs protruding from a rather reduced armour. The purely terrestrial tortoises did not appear until Tertiary times.

Ichthyosaurs

The fish-like ichthyosaurs probably developed from forms similar to the small aquatic mesosaurs which we have already met (p. 193). The oldest ichthyosaurs appeared in the middle Triassic and were already well adapted to life in the sea. They were streamlined air-breathing creatures, usually about ten feet long with long snouts, sharp teeth, large eyes, two pairs of paddle-like ventral limbs, and a fleshy dorsal fin (Plate 4*b*). In the early forms the tail was long and pointed, but in later Jurassic and Cretaceous forms it became crescentic and fish-like in outline with the vertebral column down-flexed into the lower lobe. It was the tail which was the chief agent of propulsion, the fins acting chiefly as steering paddles. Some

THE DOMINANCE OF THE REPTILES

remarkably well-preserved fossil specimens show that the ichthyosaurs gave birth to their young alive. The ichthyosaurs were one of the earliest and perhaps the most successful of the aquatic reptilian groups. They seem to have been fast and powerful swimmers in the open ocean and they persisted until Upper Cretaceous times.

Marine crocodiles and mosasaurs

If the ichthyosaurs were the most highly adapted marine reptiles, the mosasaurs and geosaurs or thallatosuchians were the most savage. The geosaurs were sea-going crocodiles which are found only in the Jurassic and Lower Cretaceous. The mosasaurs, on the other hand, are confined to the Upper Cretaceous, during which time they spread throughout the world (Fig. 31). They were marine lizards, some of them thirty feet long, with long powerful tails and jaws and the general appearance of sea serpents.

Nothosaurs, placodonts, and plesiosaurs

The three remaining groups of marine reptiles are all closely related. They moved through the water chiefly by the use of oar-like limbs; the groups we have just mentioned swam by side-to-side undulations of the body and tail. The nothosaurs were small creatures, rarely more than four feet in length and probably amphibious in habit. They were slender, long-tailed, long-necked, and long-snouted, with limbs that could apparently function as both fins and feet (Fig. 31). They were probably similar in general form to the ancestors of the plesiosaurs.

Both the nothosaurs and the placodonts are known only from the Triassic. The latter were ponderous creatures, with jaws highly modified for crushing shellfish (Fig. 31).

The plesiosaurs were a far more successful and widespread group than these. They are found throughout the Jurassic and Cretaceous and some of the later forms were fifty feet in length. They have been described as resembling a 'snake threaded through a turtle', and they developed strong, large

THE EVOLUTION OF LIFE

paddles and sharp teeth (Plate 4(*b*)). Some forms were short-necked with enormous skulls (one is ten feet long): others were long-necked (sometimes twice the length of the body) with smaller skulls.

It was these various fish-eating creatures of ocean-going power and proportions which established the reptilian rule of the seas. But, for all their prowess, they remained reptiles, and when the changing conditions at the close of Cretaceous time brought extinction to their relatives on the land, they were not immune from their touch – they vanished from the scene.

THE AIR

The three environments in which we have so far traced the dominance of the reptiles, the land, inland waters, and the seas, were not unfamiliar to the vertebrates. In these environments the reptiles consolidated the vertebrates' invasion, but they did not initiate it. There was, however, one other environment that was new – the air – the last major environment to be colonized by the vertebrates.

In this last great conquest the reptiles played a major part. We shall consider it further in Chapter 10.

THE DECLINE OF THE REPTILES

The progress of the Mesozoic saw the growing world-wide dominance of the reptiles. The end of the Mesozoic saw their rapid decline, and for the dinosaurs, plesiosaurs, ichthyosaurs, and pterodactyls it brought extinction. The end of Cretaceous times has well been described as 'the time of great dying'. Of course, it is easy to over-dramatize the suddenness of it – some dinosaurs, the stegosaurs, for example, had already become extinct in the Lower Cretaceous – but the rest of them, the savage tyrannosaurs, the giant quadruped sauropods, the duck-billed ornithopods, the armoured ankylosaurs, and the horned ceratopsians, all these became extinct towards the end of the Cretaceous, and geologically this was sudden death.

THE DOMINANCE OF THE REPTILES

How did it come about? There have been many suggestions. The reptiles were too stupid, too cold-blooded, too specialized, their race was senescent, they were plagued by disease, by genetic maladjustment, by overactivity of the pituitary glands, they were outwitted and overwhelmed by the mammals, who preyed upon their eggs – so the suggestions go. Now all these suggestions may be valid on a limited scale – they may well have been important in the local survival of some groups and some of them may even have contributed to the general extinction of the reptiles. But none of them, singly or collectively, seems to be adequate to account for this great dying. The dinosaurs were stupid – the mighty forty-ton *Diplodocus* had a brain the size of a hen's egg – but dinosaurs had survived for 150 million years, for all their stupidity – and cold blood, pituitaries, and endocrines too, for that matter. And the mammals cannot really have presented any sudden new problem – they had shared a lowly place in the reptilian world for 100 million years or so, and it was only after the extinction of most of the reptiles that they 'exploded' in a great surge of evolutionary activity. No, the answer seems to lie outside the dinosaurs themselves, outside their enemies – it seems, in fact, to lie in the changing earth on which they lived.

Upper Cretaceous times were remarkable over much of the earth for several reasons. They were marked by one of the greatest transgressions of the sea that has ever been known in the earth's long history. Huge areas of land in Europe, North America, North Africa, and elsewhere were inundated by this great spread of the Cenomanian seas. In many places new mountain ranges were coming into being (the Rockies and the Andes among them) and, with them, changing, more varied, and often colder climates. Certainly there had been earlier geographical changes than this, and the dinosaurs had survived them; but now came the end of an era of earth history and the transformation of the geographical pattern that had provided the familiar homes of the reptiles for more than a hundred million years. And with the changes in the earth itself, there went changes in its plants – the old vegetation of the Mesozoic was replaced by flowering plants and hardwood forests. For the first time in the earth's long ages, there were trees and plants barren of leaves through the cold winter months.

For the gigantic sauropods, at least, it was probably these dramatic changes that were the agents of extinction. For the great size of these creatures, their cold blood, their small brains, their specialized diet, and their probable restriction to low-lying swamps and lakes must have made them particularly vulnerable to almost any change, and especially such changes as these. If their old haunts were submerged by the rising seas, or the widespread warm temperatures and lush vegetation of the late Mesozoic were changed, these monsters would be almost incapable of migrating to new areas. With their decline, the carnivores which preyed upon them would also suffer.

It may well be then that the inability of dinosaurs to adapt themselves to the changing conditions of Upper Cretaceous times was the basis of their decline. Yet it is easy to oversimplify this explanation, for it was not only terrestrial dinosaurs, but also ocean-going ichthyosaurs, plesiosaurs, and mosasaurs and the flying pterosaurs that vanished from the scene. It is very probable that there was no single cause for this mass dying. The maintenance of life is an almost inconceivably intricate combination of many complex processes and its mass extinction is probably the result of no less intricate and numerous factors. Yet if we cannot give any adequate explanation for the decline of the ruling reptiles, we can at least suggest that it was probably connected in some way with their inadaptability to a changing earth.

Chapter 10

THE AIR

THE seas, the rivers, the land, the air – we have now seen something of the development and diversity of life in each of these realms. We have seen too that one of the characteristics of the evolution of living things has been their tendency to increase in range of adaptation and thus in numbers. The zone of life, the biosphere, has expanded since life's youth, and this expansion has involved not only the invasion of new environments, but also the fuller use and greater exploitation of every habitat in those environments already partly colonized. This is the pattern we saw on the land. The first foothold was gained by the amphibia, but they were limited by their structure to wet or humid areas. The rise of the reptiles exploited the empty upland environments. The same sort of sequence marked the invasion of the air.

The earliest aviators were the insects, and for about 100 million years they had the monopoly of the air. With increasing competition on the land, however, other creatures took wing and three distinct groups of flying vertebrates developed. Both the advantages and the limitations of life in the air are readily apparent, especially in connexion with food-gathering, protection from enemies, and rapid colonization of new environments. But before we consider the history of flying things we need to remind ourselves that the modifications involved in the transition from land to air are every bit as profound as those involved in the earlier transition from water to land. In the long history of life, the conquest of the air represents another major landmark.

THE INSECT VANGUARD

The insects are arthropods which need little introduction, for they play an important, familiar, and sometimes disastrous part in the life of man. They are the most abundant of all living groups, the most widespread of all land animals, and the only invertebrates to have conquered the air, although not all forms are capable of flight. The oldest unquestionable winged insects occur in the Upper Carboniferous, and include forms of two very distinct types. This implies a long earlier history, and indeed primitive wingless insects are known from specimens in the Devonian Rhynie Chert of Scotland. It seems probable that it was from such forms that winged insects developed.

In spite of their somewhat obscure origin, however, insects were widespread and diverse by the latter part of Palaeozoic time and, though they are seldom preserved as fossils under most conditions, a surprising number of fossil forms have been described. About half of these are known only by their more resistant wings, although under favourable conditions other structures may be preserved, and even such things as pupal cases, larval chambers, and borings in fossil wood are known.

By Upper Carboniferous times a number of familiar groups were well established: the cockroaches reached their acme of development (some were four inches long); one of the other ten orders (now extinct) was remarkably similar to our modern dragon-flies, and included giant carnivorous forms with a wing span of about thirty inches. Most of the other groups were archaic, many broadly resembling grasshoppers, but by the end of the era a number of modern groups had made their appearance – lace-wings, stoneflies, mayflies, cicadas, beetles, and thrips. In one sense, in fact, the insects have not undergone any fundamental change since the appearance of Permian forms capable of complete metamorphosis.

Many other familiar forms are only a little less ancient: true bugs and flies, earwigs, ants, and caddis flies all appear in the Mesozoic, and butterflies, moths, and other social insects (bees, etc.) in the Cenozoic.

But the success of the insects as the 'first aviators' is only a partial measure of their achievement. They are the most abundant of all living groups. In numbers of species they outnumber

THE AIR

all other living animals combined by about three to one. In terms of abundance of individuals they are equally prolific. They multiply with astonishing rapidity under favourable conditions. The offspring of a single aphid could cover the earth in one season if all survived! They are highly adapted to life in all environments except the oceans (although some are littoral and one form lives on the surface of the sea), and they have a wider distribution than any other group of land animals. Some forms are parasitic, others live in ponds and streams: some are known to migrate to new feeding areas over distances of several hundred miles. Many insects are solitary, others live a social existence in highly organized societies. Insects play a significant role in the life of mankind. Some are important food producers, such as bees and other insects which not only produce honey but also play an essential part in the cross-fertilization of many fruits and crops. Others are pests and carriers of disease against which man is obliged to fight a continuous battle. Indeed, in many ways the insects are man's greatest competitor on the earth.

FLYING REPTILES

The oldest known flying vertebrates come from the Jurassic, and, as if to emphasize the invasion of the air, two distinct but related groups are represented. The first group to appear, the pterosaurs ('winged lizards') remained reptiles. The other group were reptiles with feathers – the ancestral birds. Both sprang from the thecodont reptiles. A 'gliding reptilian' fossil (a protorosaur) has recently been reported from the Trias of New Jersey, but it has not yet been fully described.

The pterosaurs showed three essential modifications for life, in the air – the presence of wings, a light but strong skeleton, and large cerebral hemispheres which control the important senses of sight and coordination. Thus many of the bones were hollow and air-filled, the skeleton was given added rigidity by the fusion of certain parts, and some structures (such as the sternum or breast bone) were enlarged. Other

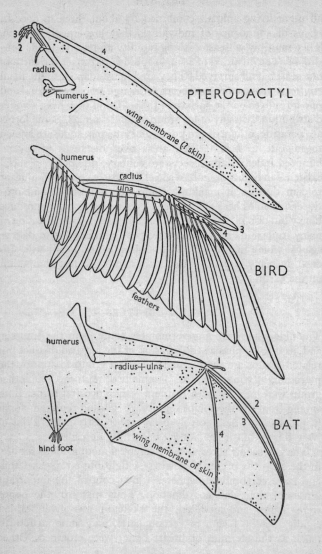

THE AIR

structures were modified, such as the fourth finger, which formed an elongated support for the broad membranous wings, the remaining three fingers protruding as claws from the leading edge (Fig. 35). The Jurassic pterosaurs were rather small creatures (some no larger than a sparrow) with, in some cases, toothed jaws and long spade-like tails, but the Cretaceous forms were much larger – some (*Pteranodon*) with a wing span of twenty-five feet and long pointed 'hammer heads' (Fig. 36B). The toothless skull was nearly twice as long as the vertebral column.

The general construction of the pterosaurs suggests that they were less active fliers than are most birds, and they probably supported themselves by extended soaring and gliding. Although fossil walking tracks have been found they may well have been largely arboreal in habit, for the claws of some forms were well adapted for clinging to rocks and trees. Their wing structure was less robust and less efficient than that of birds or bats (Fig. 35), and if they shared the cold-blooded existence of other reptiles (though this is far from certain) they must have been severely limited in their level of activity.

Most of their remains are found in marine strata, and these creatures were probably fish-eating sea fliers, though others apparently lived on land, perhaps feeding on insects or fruit. Their delicate skeletons would make preservation as fossils unlikely, although a few fossil forms are known from fresh-water deposits.

The pterosaurs lived throughout the 100-odd million years of Jurassic and Cretaceous time, but they shared the fate of their cousins as the Mesozoic drew to a close. For all their external difference they remained reptiles, and they too were caught up in the great dying at the end of the Cretaceous.

FIG. 34 Comparison of wings of pterosaur, bird, and bat. The fingers are numbered 1–5. In the reptile the membranous wing was supported by the elongated fourth finger, the other fingers being hook-like claws. In birds the fingers are combined to give added strength, and the wing is formed of stout primary feathers. In bats four fingers are extended to support the wing membrane. (*After Storer*)

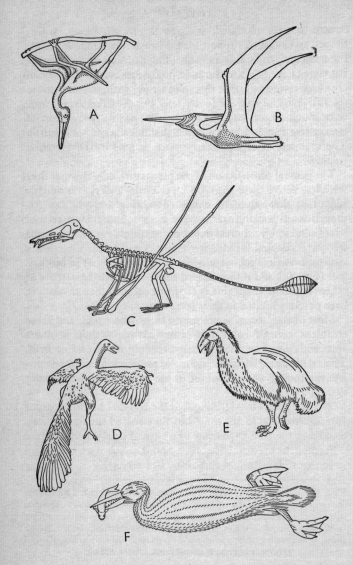

THE AIR

BIRDS

The birds (Class *Aves*) include some of the most familiar and distinctive members of the animal kingdom. However hazy our notions of the characteristics of other groups, none of us needs to be reminded that birds are covered with feathers, that they possess wings, that their skeleton is of delicate but strong construction, that they have beaks but no teeth, and that they lay eggs. Clearly, in many of these respects they stand alone from other animal groups. What is rather less obvious, however, is that the level of their body structure is 'higher' than all other animals except the mammals. Thus the feathers provide an insulating covering, their body temperature is regulated (they are 'warm-blooded') and their circulatory system highy specialized, their ability to fly allows them to occupy unique environments, their senses of hearing and vision are highly developed, and their song has been a source of delight to mankind down the ages. But perhaps most important of all is the relationship between the parent birds and their offspring. Many birds are born in a naked, helpless condition, others are born fully feathered and able to 'stand on their own feet', yet in both cases the young fledgelings are nurtured and cared for by their parents. Just as the reptiles' reproduction marked the consolidation of life on the land, so also this distinctive feature of avian reproduction marked the initiation of family and ultimately society relationships which have played so fundamental a role in the life of our own and other species.

FIG. 36 Flying reptiles and birds.
A Sketch of pterosaur, showing probable method of walking; B *Pteranodon*, a Cretaceous flying reptile, with a wing span of 27 feet; C skeleton of a Jurassic pterosaur, *Rhamphorhyncus*, showing greatly extended fourth finger and rudder-like tail; D *Archaeopteryx*, a Jurassic toothed bird, about the size of a crow; E *Diatryma* a flightless bird, 7 feet tall, from the Eocene of Wyoming; F *Hesperornis*, a Cretaceous diving bird, 6 feet in length. (*A after Abel, B after Seyfarth, C after Seeley, D after Seyfarth, E after Hayes, F after Gleeson*)

The birds are such accepted members of the world in which we live that it is difficult to imagine a period of earth history in which they were non-existent. But they are comparative newcomers to the geological scene.

The oldest fossil birds come from the Jurassic Solenhofen limestone of Bavaria. They are represented by four specimens – a single feather, and three well-preserved skeletons, once thought to represent two genera, but all now regarded as *Archaeopteryx* (Fig. 36D). A single feather is known from beds of equivalent age in N. E. Spain. We have already seen something of *Archaeopteryx* and the conditions under which it was preserved (p. 56). *Archaeopteryx* is a jumble of both primitive reptilian and advanced bird characters. It has feathers, but also teeth. It has bird-like feet, but reptilian vertebrae and tail. It has wings, but with claws projecting from them. It has a wishbone (found only in birds), but a reptilian brain. This mixture of reptilian and avian characters gives a clear pointer to the ancestry of the birds, which appear to have arisen from small arboreal thecodont reptiles.

No specimens of earlier transitional forms have yet been found; nor is this surprising, for their probable environment and habits would make them unlikely to be preserved as fossils. Indeed, after the brief glimpse of Jurassic birds in the Solenhofen limestone, there is a gap of about fifty million years in their history, the next fossils known being those from the Cretaceous Niobrara Chalk of Kansas. Other fragmentary fossils of ancestral ducks and pelicans are known elsewhere. Such a gap is a reminder that our explanation of the absence of many 'missing links' involves no special pleading – the fossil record is every bit as fortuitous and incomplete as palaeontologists assert. The chalk of Kansas contains several genera of at least two distinct types of extinct sea-birds, and they indicate the diversity which the birds had achieved by this time. One of them, *Ichthyornis*, was a strong flier, probably rather tern-like in its general appearance and habits. The contemporary *Hesperornis* (Fig. 36F) was quite different – toothed, about six feet long, a diver-like bird, clearly a powerful swim-

THE AIR

mer – but flightless! So, after the long ascent to the air, some birds soon abandoned the power of flight, just as some amphibia cumbrously adapted themselves to life on the land, only to return to the water, and some reptiles, having strengthened the hold on the land, returned to the seas. Such is the diversity and range of adaptation – and such also its irony!

Unlike the flying reptiles, the birds survived the end of the Mesozoic, but their subsequent history is rather patchily recorded in the fossil record. Such birds as are known have a distinctively modern appearance. Teeth are no longer present and, unlike the pterosaurs, the hind legs are usually well developed. Amongst the most spectacular Cenozoic birds were the giant flightless birds, the forebears of the living ostriches, cassowaries, emus, and rheas. Some extinct forms were very like these tall wingless creatures, and the structure was apparently a response to conditions of plentiful food supply and absence of predatory enemies. Even in early Cenozoic times such forms existed; some seem to have been savage carnivores and they may well have been formidable local rivals to the mammals (Fig. 36E).

Other Cenozoic birds became adapted to widely different kinds of life, such as the penguins of Antarctic regions, waders of the shores and estuaries, sea-birds of the open oceans, song birds of woodland and meadow, game-birds of upland areas, and powerful birds of prey of the mountains and other areas. Birds vary enormously in size, from the extinct ten-foot-high elephant bird of Madagascar to some tiny humming birds which weigh only $\frac{1}{10}$oz. Some birds (pelicans and cormorants) are voiceless, others have characteristic songs of great complexity; many birds are solitary, others live in flocks of up to 100,000; some birds fly at speeds of over fifty miles per hour, others are flightless; some are restricted to one locality, others undertake short seasonal migrations, while others travel several thousand miles. Judged by any such standards, the birds are not only the champions of the air, but also one of the most successful of all living groups of animals.

THE EVOLUTION OF LIFE

FLYING MAMMALS

The only mammals capable of sustained flight are the bats, although both 'flying lemurs and squirrels' have wing-like membranes which enable them to glide. Bats are generally small creatures, with membranous wings supported by the outstretched fingers of their long fore-arms (Fig. 35) and extending to the shorter hind legs, with one or two fingers commonly protruding as claws along the edge of the wing. This wing structure is similar to that of the pterosaurs, but the incorporation of four fingers instead of one to support it makes it more efficient. One group of bats (e.g. the flying fox of Java, with a wing span of five feet) live chiefly on fruit, but most are insectivores. Bats are of world-wide distribution, and often occur in great numbers, but their habits (like the birds) make them poorly known as fossils. The oldest bats are of Eocene age, but these already have fully developed wings, and are not very unlike living representatives. They presumably arose from some group of arboreal insectivores.

Chapter 11

MODERN PLANTS

It is temptingly easy to regard the history of life as synonymous with the history of animals. The great majority of books on the subject do just this, and most of the remainder contain little more than a few off-hand references to 'parallel changes in the plants'. Yet to do this is to fail to convey the intimate connexion which exists between the two groups. The changes in Cenozoic mammals, for example, can be understood only in terms of the profound changes which were taking place in contemporary plants. Before we turn to the mammals we must consider these changes in the plants.

The prevailing aridity that marked the early Mesozoic era in many parts of the earth is reflected in the general rarity of Triassic plant fossils. The few that are known, however, show clearly the general characteristics of these floras. In the Southern Hemisphere a distinctive (*Dicroidium*) flora persisted. In northern lands a number of familiar Palaeozoic forms were still present: some cordaites, seed ferns, and scale trees survived for a short time, there were scouring rushes, and conifers were a major component of the floras. The ferns were equally important and the main groups represented had first appeared in the Carboniferous, although many genera were new. The most striking newcomers, however, were the now extinct cycadeoids (or bennettiales). These forms were broadly similar to the living cycads; they had rather short, stumpy, generally unbranched trunks, crowned by clusters of palm-like fronds. They are common fossils in the Jurassic Estuarine Series of the Yorkshire Coast. They differed from living cycads in their method of reproduction, for both the male and female reproductive organs were combined together to form a flower-like structure. For all their superficial similarity, however, these structures were really pseudo-flowers and they do not appear

THE EVOLUTION OF LIFE

to have given rise to the true flowers of later plants (Fig. 37). The cycadeoids are amongst the most widespread and abundant, and, with the ferns and the conifers, the most characteristic Mesozoic plants. Unlike the two latter groups, however,

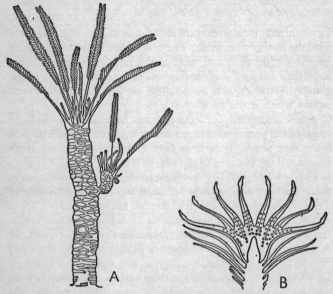

FIG. 37 Cycadeoids – the most typical of early Mesozoic plants. A An extinct cycadeoid, *Williamsonia* from the Jurassic, about 6 feet in height; B a so-called cycadeoid 'flower', in median section. (*A after Sahni, B after Wieland*)

they became extinct before the end of Cretaceous times. Their decline shows some correlation with the rise of the flowering plants, and the two events may well be connected.

The cycads still survive in warm areas, and in them the male and female cells are produced in separate cone-like structures. The cycads appeared at roughly the same time as the cycadeoids, although they were much less common in Mesozoic times.

MODERN PLANTS

Most of these Mesozoic plants were small in size, but there are also records of forests of much larger trees – the petrified forest of Arizona of Triassic (Chinle) age includes some trees ten feet in diameter and over 100 feet in height. These are forerunners of the monkey-puzzle conifers, and are exquisitely preserved in red and yellow chalcedony.

The floras of the Jurassic were varied and abundant. Forests of conifers, cycadeoids, and ferns show a general similarity in such widely scattered areas as the Arctic, Japan, the United Kingdom, and the Antarctic. This resemblance is far from complete, however, and there seems little evidence for the 'uniform Jurassic climates' of which so much has been written. The conifers of the Northern and Southern Hemispheres, for example, belong to different families. In the north ginkgos were abundant, but cycadeoids were generally absent, being largely confined to a more southerly belt.

The ginkgos were widespread and the present ginkgo (restricted to China and re-introduced to other areas) is a veritable living fossil, one of the oldest of all living trees. In spite of their fern-like foliage, the ginkgos are true seed-bearing plants, with several distinctive features. They share with the cycads, for example, the active sperms found only in such primitive groups as ferns and mosses.

Another important group of Jurassic plants are the Caytoniales, once thought to be primitive angiosperms, but now commonly regarded as more closely related to the ancient seed ferns. If this identification is correct, they represent the last members of the great group of plants which had dominated the Carboniferous coal forests.

Cycadeoids, ferns, and conifers – these are the dominant plants of the early Mesozoic. But the Jurassic also contains the first record of a group of plants with whose later development the whole pattern of Mesozoic vegetation changes. These are the angiosperms or flowering plants, whose earliest remains are represented by questionable pollen grains found in coal of Jurassic age from Brora, Scotland.

THE EVOLUTION OF LIFE

FLOWERING PLANTS

Now the groups of seed-bearing plants which we have so far considered, seed ferns, cycadeoids, cycads, gingkos, cordaites, and conifers, differ from those that follow in that the seeds are borne on structures (carpels) which do not completely enclose them. The groups are often therefore grouped together as the gymnosperms ('naked seeds').

In the angiosperms ('capsule seed plants'), which include all the flowering plants, the most distinctive feature is the protective covering or ovary that encloses the seeds. It is this feature which has made the angiosperms the most successful of all plants. Some are minute plants, only a fraction of an inch in height, others such as the giant redwoods are 350 feet high and several thousand years old. They are found throughout the earth, in every land environment from the tropics to the polar wastes and in every climatic zone from deserts to monsoon jungles. They far outnumber all other plants combined in their total number of both species and individuals, and their level of organization is more complex than that of any other group. Almost all plants play a part in the life of man, but we are particularly dependent upon the angiosperms. They make up most of our vegetable diet (fruit, vegetables, cereals, and so on) and most of the animals bred for meat exist also upon them. Angiosperm trees supply a large share of our timber supplies, and their flowers provide an endless source of joy.

Angiosperms may be divided into two groups. The monocotyledons develop from a single leaflet, and are undifferentiated into pith, wood, or bark, having no growth rings. The lilies, palms, onion, bananas, and grasses are familiar members. In these the roots are fibrous, the leaves generally parallel veined, and the flower parts usually developed in groups of three.

The dicotyledon seedling has two leaflets, the stem is differentiated into pith, wood, and bark, and annual growth rings are formed. Almost all the familiar garden plants, (except tulips, orchids, and lilies), the deciduous trees and

MODERN PLANTS

shrubs, fruits and vegetables fall within this group, in which the leaf veining is usually reticulate, and the flower parts have a four- or five-fold grouping.

Angiosperms first became widespread in Cretaceous times. In the floras of the Lower Cretaceous angiosperms are either absent or very rare, although probable remains are known from strata as old as middle Triassic. By the end of the Cretaceous they are represented by more than ninety per cent of all known species, in contrast to a combined total of only five per cent for the once ubiquitous cycads and conifers. That is the measure of the angiosperms' dominance. In the long history of life there are few comparable examples of such rapid evolution.

By the close of the Cretaceous several hundred species of flowering plants were in existence. Some of these were primitive forms. Others were the earliest representatives of familiar genera which still survive – beech, birch, maple, oak, walnut, fig, magnolia, poplar, and willow trees, as well as holly, ivy, laurel, vines, and other plants and shrubs. Temperate conditions apparently existed far to the north of their present limits, for a large number of typical temperate genera occur in Cretaceous strata of Arctic regions.

It may well be that the spread of the angiosperms was favoured both by increasing aridity in some areas in late Jurassic to early Cretaceous times and by the great mid-Cretaceous (Cenomanian) marine transgression which brought extinction to many floras. When the seas retreated towards the close of the period, it was the angiosperms that colonized the empty lands.

This transformation in the plant world was of the greatest importance, not only for the vegetable kingdom but also for the animals, for the appearance of plentiful supplies of edible nuts, fruit, seeds, and leaves was a major factor in the rapid contemporary evolution of birds and mammals, and also possibly the flying reptiles. The development of flowers implies the spread of pollinating insects, though little is known about them as fossils.

The conifers remained dependent on the wind as the chief agent for seed dispersal, but many angiosperms became adapted

THE EVOLUTION OF LIFE

to dispersal by other means, chiefly birds, insects, and mammals. This has probably been an important factor in their success, and it is yet another example of evolutionary interdependence.

What a contrast these changes must have brought to the face of the earth. When the Mesozoic dawned, the earth was clothed with conifers, ferns, cycads, and a few relics of the coal-forests: a sombre landscape of greens and browns. Reptiles spread across the land and only the archaic insects rose through the heavy air. There is a sense in which the changes of mid-Mesozoic times came as a worldwide spring in the long life of the earth. Flowers, butterflies, birds – these were tokens of a new age.

The cycadeoids became extinct during the Cretaceous. The cycads and conifers still survive, but both groups are very much reduced. The cycads, although widespread, are confined to a relatively few genera in warm or sub-tropical areas. The conifers, for all their mighty forests, are also a limited group. They tend to be confined to the less hospitable, cold, upland environments of the earth, and are represented by comparatively few species. They had been the characteristic flora for a period of 75 million years, but their decline began with the blossoming of the angiosperms. For in plants, no less than in animals, the principle of natural selection applies – it is the best-adapted that survive.

We have seen something of the far-reaching implications of the rise of the angiosperms, not only in the general change that they brought to the face of the earth, but also in the related changes that took place in insects, mammals, and other plants. There is one other group of animals in whose decline the angiosperms are sometimes said to have been a major factor. The dinosaurs had spread and multiplied throughout the Mesozoic, in a balanced economy that was at once both successful and precarious. In Upper Cretaceous times they vanished from the scene. We have already speculated about the factors involved in their sudden extinction, and have seen the difficulty of any simple explanation. Yet it has been suggested

MODERN PLANTS

that the great plant changes of Cretaceous times were influential factors in the decline of the giant herbivores and thus, indirectly, in that of the carnivores. Whether or not this is so is problematical, for some herbivorous dinosaurs underwent a considerable expansion in Cretaceous times. It may be true, however, of the huge sauropods, which probably obtained most of their food from the lakes and swamps in which they lived.

CENOZOIC PLANTS

To write of Cenozoic plants is to risk confusion, for the Cenozoic, the era of modern life, is based upon characteristic fossil animals. Had it been based on fossil plants (the Cenophytic), it would presumably begin in Triassic or Cretaceous times (according to the predilections of the founder). But it nevertheless remains true that post-Cretaceous plants are different in some respects to those of the Mesozoic. Most of these differences are rather minor, however, and are chiefly represented by the incoming of modern species. Many hundreds of species of Tertiary plants are known, and members of almost all genera show an increasing similarity to forms still living. Because of this similarity, the abundance of fossils, and the relatively short period of time involved, these fossil floras provide an unusually good opportunity to compare the distribution of present floral zones with those of the past. If we assume that the fossil species involved occupied broadly similar environments to those of their descendants, it becomes possible to reconstruct ancient floral zones. The general result is one of permanence, with change: permanence, because floral zones undoubtedly existed in the past and appear to have been parallel to those which still exist today; change, because with changing climates the plants which constitute the zones have constantly migrated back and forth. London was the site of great tropical forests in early Tertiary times, and the remains of the ancient plants are preserved in countless numbers in the London Clay. In the Pacific north-western states of Oregon and Washington, the

THE EVOLUTION OF LIFE

Tertiary Eocene strata yield abundant palms and other plants which today grow only in the wet and warm areas of Central America and southern Mexico. Beds of the same age in Alaska contain what are still more typical northern plants, maples, sequoias, poplars, willows, and birches. In the overlying Miocene rocks of Washington and Oregon these dryer, colder forms have replaced those of the Eocene, the more luxuriant fauna having moved southwards.

Perhaps nowhere is this floral migration better shown than in the Pleistocene glaciation, when ice sheets of continental proportions advanced and retreated across the Northern Hemisphere in a series of climatic fluctuations. Very clear evidence is provided of glacial conditions gradually giving rise to interglacial periods in which varying groups of plants re-established themselves.

So far we have traced two aspects of Cenozoic floras – their increasing modernization and their changing distribution. One further factor remains – the spread of the grasses. Grasses are unlikely candidates for preservation as fossils. Occasional and mostly unauthenticated records of them exist in Upper Cretaceous and early Tertiary rocks, but they are generally absent. Recent studies suggest that they first became abundant in Miocene times, for their seeds are known in great numbers from the Miocene strata of the High Plains of Kansas. Their common appearance has a double interest. Firstly, it is an indication of widespread physical changes which brought about a general transition from forest conditions to open prairies. Secondly, their appearance coincides with very rapid and profound changes in the mammals. These great changes (p. 241) concern many groups and seem to point clearly to a general change from browsing to grazing habits. There can be little doubt that the changes are very closely related. To describe them as cause and effect may perhaps be an oversimplification – but it is not far wide of the mark.

Chapter 12

THE RISE OF THE MAMMALS

EARLY MAMMALS

THE mammals (Class *Mammalia*) include the great group of living things that most of us refer to simply as 'animals'. This is a measure both of their familiarity to us, and of our recognition of them as the 'highest' forms of life. The Class includes most of the common domestic and farm animals as well as many of the more spectacular zoo animals, the bats, and the whales. All the mammals are more or less covered with fur or hair, are warm-blooded, have highly developed senses, and almost all of them give birth to their young alive and nourish them with milk secreted from the mammary glands of the mother. This parental care is practised to varying degrees in various forms, and reaches its peak in man. There are also various distinctive skeletal features, such as the strongly differentiated teeth and the structure of the ear and the jaw. There is one other mammalian feature, less diagnostic than some others, but yet the ultimate key to their dominance – that is their brain size. In no other living creatures is there such striking mental development. Although they are relative newcomers, the mammals are one of the most successful of all living groups. They exist in virtually every part of the earth from the poles to the tropics, from the depths of the seas to the air, and they range in size from the mouse to the blue whale, the largest animal which has ever existed, reaching a length of 120 feet and weighing 150 tons.

The first mammals appeared in the late Triassic, and are represented by the fossil teeth and jaws of very small creatures, most of them about the size of mice and shrews, and probably similar in general appearance. Their immediate ancestors are unknown, but it seems almost certain that they arose from forms resembling the theriodont or ictidosaurian reptiles (p. 196). There are

THE EVOLUTION OF LIFE

four orders represented in the Jurassic, each based on a characteristic tooth structure. Two of these (the symmetrodonts and the pantotheres) are confined to the Jurassic, the latter apparently being the ancestors of the later mammals. One of the two other groups, the multituberculates developed into a variety of herbivorous forms, some of which survived into the Early Cenozoic. The remaining triconodonts were cat-sized carnivores and survived to Cretaceous times.

It may be that other Jurassic mammals existed, but we have no records of them. The living monotremes, the duck-billed platypus and the spiny anteaters of Australia, for example, are very primitive animals: they lay eggs, they have some reptilian skeletal characters and they secrete milk from modified sweat glands, which are homologous to the breasts of other mammals. Although they appear as fossils only in the Pleistocene, it seems probable that they arose much earlier.

Living marsupials are a relatively minor group of mammals, being almost insignificant in comparison with the hosts of their placental brothers. They are represented by kangaroos and various other animals of Australia, and the opossums of South and North America. In these creatures the young are born as tiny immature embryos, which are then sheltered in the mother's pouch, where they attach themselves to the milk teats by which they are fed. Many features of the marsupial skeleton are primitive.

The oldest marsupials are opossum-like creatures from the Upper Cretaceous of North America. From these forms there evolved a great variety of Tertiary descendants, including South American sabre-toothed forms as big as tigers. South American Tertiary faunas contain many marsupials, but almost all of them became extinct when the Panama isthmus was re-established in later Tertiary times, providing a bridge by which the more advanced placental North American mammals were able to invade South America. The decline of marsupials in the continent appears to be the result of competition with these creatures (see p. 246).

In Australia a large number of marsupials still survive –

THE RISE OF THE MAMMALS

among them the carnivorous Tasmanian wolf, the kangaroo, the wallaby, the tiny springing bandicoot, the koala bear, and the wombat. The isolation of Australia has provided a sheltered environment, free from placental mammalian competition, and it is to this that both the marsupials and the monotremes owe their survival. Their adaptation to various environments shows many striking (but quite independent) parallels to the placental mammals.

PLACENTAL MAMMALS

We have so far considered two groups of living mammals, the primitive monotremes and the pouched marsupials. All other living forms (by far the greatest group) belong to a third order, the placentals (or Eutheria), characterized by the development of a structure known as the placenta, by which the developing embryo is attached to the wall of the mother's uterus and is fed by oxygen and food from her body. This method of reproduction results in the birth of young at a much more advanced stage than in marsupials, although some of the marsupials also have placenta-like structures. This longer prenatal development and the period of parental care and training that follow it have been features of the greatest importance in the growing dominance of the placental mammals. Perhaps the most spectacular example is afforded by the blue whale, whose young are twenty-three feet long at birth and fifty-two feet at weaning! Equally important has been the fact that most placentals have relatively much larger brains than any other group. There are also various other distinctive features by means of which fossil forms can be readily recognized.

The oldest placentals appeared in the Cretaceous, at about the same time as the marsupials, and it seems probable that both arose from a common ancestor. The early placentals were small insectivores – the forerunners of living shrews, moles, and hedgehogs – to which some forms were broadly similar, though others were more primitive in appearance. All were small, and probably unobtrusive, nocturnal creatures, totally

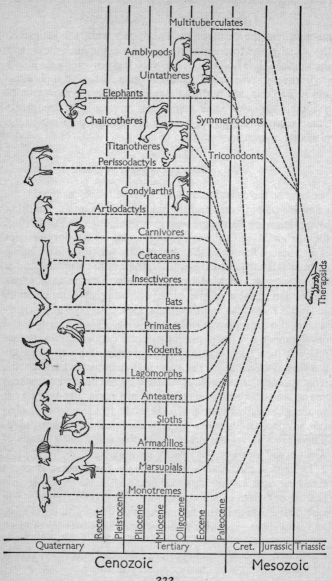

THE RISE OF THE MAMMALS

insignificant in comparison with both the dinosaurian giants amongst whom they lived, and the great galaxy of mammals that developed from them.

This then is the record of the early mammals, an obscure and lowly group throughout Jurassic and Cretaceous times. Their day was not yet, but it was to come and with its coming there dawned the last great era in the history of life – the Cenozoic – the Age of Mammals.

FIG. 38 The evolution of the mammals. (*After various authors, especially Scheele and Colbert*)

Chapter 13

THE AGE OF MAMMALS

The close of the Mesozoic saw the extinction of the ruling reptiles, which for well over 100 million years had lived as undisputed masters in every environment on the land, in lakes, rivers, in the seas, and in the air. With their passing, once crowded areas became empty, and after almost 100 million years of insignificance and subservience, the mammals spread across the earth to establish a new supremacy.

We have already seen that by Cretaceous times both the marsupials and the placentals were well established. Both groups still survive, but it was the placentals that achieved the mammalian conquest. It was they who filtered into the empty lands and seas of the Cenozoic and produced the latest race of champions.

For the first time in the long history of life, many of these creatures will be familiar to us, for the 70-odd million years of the Cenozoic era include the moment of time in which we dwell today. The same broad similarity applies to the position and outlines of the continents, although Asia and North America were periodically joined at the Bering Straits and North and South America at the Panama isthmus. Almost all the world's mountain chains were already formed. This is not to imply that conditions became static: on the contrary, we shall see something of the subtle interaction between mammalian migration and evolution, periodic continental uplift and the gradually changing climates which culminated in the recent glaciation of the Northern Hemisphere.

ARCHAIC MAMMALS

The opening of the Cenozoic saw the rapid differentiation of two main groups of mammals, the hoofed mammals and the carnivores, both of which seem to have arisen from primitive

THE AGE OF MAMMALS

insectivore ancestors. The pattern of their development was far from uniform throughout the world, as we shall later see, but the Northern Hemisphere and Africa shared a broadly similar fauna.

The carnivores are characterized by their tearing and biting teeth, supple limbs, and claws. The earliest flesh-eaters were the creodonts, an archaic group now extinct. They showed a great diversity of form and size, ranging from weasel-like creatures to savage beasts as large as lions. Their teeth were little modified to the characteristic dagger-scissor pattern of modern carnivores, and their brains were rather small, but they were still a remarkably successful group, although most of them did not survive after Eocene times.

The primitive ungulates presumably provided the creodonts' chief prey. The ungulates are hoofed, herbivorous mammals, and include such modern forms as horses, cattle, elephants, deer, camels, and hippopotami. Their earliest members, the condylarths, appeared in Paleocene times.

The condylarths underwent rapid changes, especially in their limbs and in their teeth, which became adapted to the chopping, munching, grinding action characteristic of herbivores. The modification of the limbs is equally characteristic. The earliest ungulates had claws, but these were rapidly transformed into hooves, while the limb bones tended to become longer. It was these changes that have given many modern ungulates (horse, deer, etc.) such qualities of speed. *Phenacodus*, a goat-sized creature, is typical of these early forms; in spite of its long carnivore-like body and tail, it already showed typical ungulate features in its broad cheek teeth and its tiny hooves. Many of the descendants of these forms grew to a considerable size. The amblypods ('slow-footed' – a most expressive name; Fig. 39B) included the pantodonts, up to three or four feet high at the shoulder, with broad feet and most unherbivore-like, dagger canine teeth, and the contemporary uintatheres (Fig. 39A), which were rather larger, rhinoceros-sized animals, with elephantine limbs and six horny protuberances projecting from their massive heads. They became extinct in Late Oligocene times.

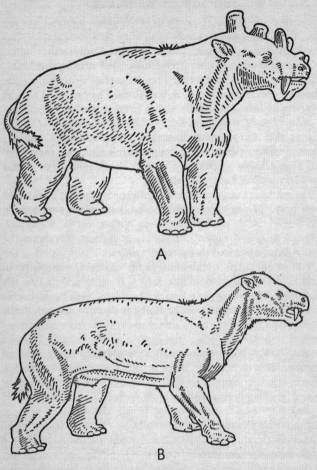

FIG. 39 Two ancient hoofed mammals from the Eocene. A *Uintatherium*, a uintathere about the size of a rhinoceros; B *Coryphodon*, an amblypod (*after Colbert*)

THE AGE OF MAMMALS

THE ARRIVAL OF MODERN MAMMALS

The second great period of Cenozoic mammalian evolution took place in Upper Eocene and Oligocene times when the archaic carnivores and ungulates were replaced by essentially modern forms.

The carnivores were represented by the fissipedes (split-feet flesh-eaters, and opposed to the web-footed marine forms), a group which includes the living cats, hyaenas, dogs, bears, and weasels. The early fissipedes were rather small, slender creatures, but already they represented a considerable increase in brain size over the creodonts. From these ancestors the modern carnivore groups arose at various times throughout the Tertiary. The cats, for example, appeared in the Oligocene and developed not only into lithe active forms represented by modern lions, tigers, and leopards but also into the spectacular 'sabre tooth' members of their tribe, which were widespread throughout much of Tertiary time. Their savage dagger-like upper canine teeth were about seven inches long in some forms and seem to have been an adaptation which enabled their owners to prey on the many thick-skinned herbivores of their day. They were undoubtedly capable of inflicting a severe wound. One skull of the cat *Nimravus* has been found which has a partly healed wound, inflicted presumably when the creature was young but healed over by the time it died in young adulthood. The shape and size of the wound clearly show it to have been inflicted by the sabre-toothed cat *Eusmilus*.

Dogs appeared in Oligocene times as relatively small creatures, which later gave rise to the more modern foxes, wolves, dogs, and coyotes. The bears probably arose from dog-like ancestors in Miocene times, most of their members becoming omnivorous feeders.

The later Cenozoic history of all these groups is one of increasing modernization. By Pleistocene times virtually all living species, as well as a number of recently extinct forms, were already in existence (Fig. 38).

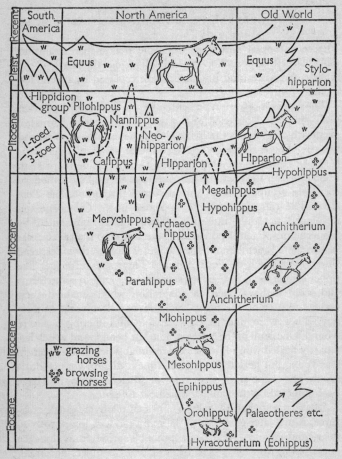

FIG. 40 The evolution of the horses.
The main lineages of the horse family showing the development of one-toed grazers and the geological and geographical distribution of the family. The restorations are to scale. (*After Simpson*)

THE AGE OF MAMMALS

Modern ungulates are represented by two more or less distinct groups: the odd-toed horses, rhinos, and tapirs (perissodactyls), and the even-toed, cloven-hoofed pigs, deer, hippos, and cattle (artiodactyls). Both groups appeared in the Eocene, and their history is well recorded in the fossil record.

There is probably no more widely known example of evolution than that of the horse. The earliest horses ('*Eohippus*' or *Hyracotherium*) were small, swift creatures, with about the size and the build of a whippet, and with four toes on the fore feet and three on the hind. The subsequent history of the horses is summarized in Fig. 40. We can observe four main changes in this classic evolutionary sequence, involving an overall increase in size, lengthening of the limbs and modification of the feet, changes in the skull and teeth, and an increase in both relative size and complexity of the brain. Although the details of this broad process of development are complex and there has been a tendency in the past to oversimplify the pattern of horse evolution, it is useful to discuss these four trends, for they provide an exposition of the evolutionary process.

The general tendency towards increase in body size was a common one in Cenozoic mammals. *Eohippus* stood only twelve inches high at the shoulder; the modern carthorse may stand over five feet (fifteen hands). Such increase in size clearly confers considerable advantages on the animals involved, but it also presents problems. An increase in height of four times, for example, involves a relative increase of sixty-four times in weight, but a corresponding increase of only sixteen times in the strength of the limbs (based on their area of cross-section). Similar problems are involved in feeding. Clearly, therefore, we should not expect present-day horses to be simply larger editions of *Eohippus*, for such (hypothetical) animals would be mechanically unsound. What we do find, in fact, is that the whole animal has undergone quite drastic change in structure, much of which is an indirect, though not always obvious, response to overall increase in size. Changes in proportion of limbs and teeth (Fig. 41) show this trend quite well.

But we cannot explain all the changes in horse structure on

BROWSERS		GRAZERS	
Pad-footed	Teeth low crowned with NO cement	Spring-footed	Teeth high crowned with cement
4-toed	3-toed	3-toed	1-toed
Pleistocene			Equus / Equus / Equus
Pliocene	Hypohippus / Hypohippus	Neohipparion / Hipparion / Hipparion	Pliohippus / Pliohippus / Pliohippus
Miocene		Merychippus / Merychippus / Merychippus	
Oligocene	Mesohippus / Mesohippus / Mesohippus		
Eocene	Eohippus		

240

this simple basis of increase in size. The teeth, for example, not only gradually increased in surface area, length (and therefore life), and efficiency of structure – all changes we might associate with general increase in body size: they also underwent a rapid modification in Miocene times which appears to reflect a radical change in the animal's habits from browsing to grazing (Figs. 41 and 43).

The same is true of the limbs. The changes in length and proportion of the limbs were largely associated with the changes in bulk of the animal, but with these changes went a reduction in the number of functional toes (Fig. 41) which reflects a complete change from a 'pad-footed' to a 'spring-hoofed' posture.

Nor was the expansion of the brain or the skull a simple increase in size. The brain increased both in size and complexity, as is shown by the relatively much greater size increase and 'wrinkling' of the neocortex (Fig. 42). The skull too became modified in form, in ways independent of tooth and brain changes. The position of the eye gradually moved towards the back of the skull, and this was probably a marked advantage to an animal that spent most of its time cropping grass.

Such a sequence of intergrading fossils is a clear confirmation of both the continuity and the complexity of the evolutionary process. There appears to have been no overall simple pattern or direction of evolution in the horses, and no constant trends or rate of change within the various groups. The one broad relationship that does emerge, however, is that between adaptation and evolutionary change. The biggest single transition in the horses is the rapid change from browsing to grazing habits, which took place in one group in Miocene times, and

---

FIG. 41 The evolution of the horses – changes in the limbs, skull and teeth.

Stages in the evolution of the fore feet of horses – oldest below – not drawn to scale. Each vertical column represents a distinctive mechanical type. The teeth shown are the grinding surfaces of right upper molars, all drawn to same scale. The skulls are also drawn to scale. (*After Simpson*)

THE EVOLUTION OF LIFE

which is reflected in limbs, feet, teeth, and skull. The development of deeper, strongly ridged teeth, strengthened by the

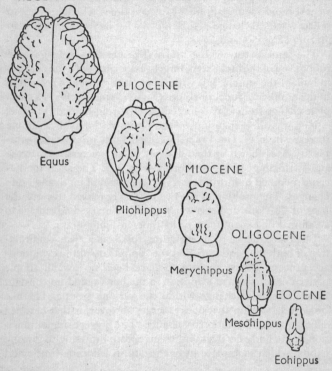

FIG. 42 The evolution of the horses – changes in the brain.
A series of casts of the inside of the cranial cavities of selected genera of fossil horses, drawn to scale. Oldest lower right. Note the relative increase in size and convolution of the cerebrum. (*After Simpson*)

growth of cement, was clearly an adaptation to feeding on the harsh grass of the open prairies, just as changes in limbs and feet seem to have fitted their owners for this new life of the

THE AGE OF MAMMALS

plains. The single hoof, for example, would have been ill-adapted for movement in the scrub of forests, but admirably suited for running on the firmer ground of the prairies. It is no coincidence that this transition from forest browsing to prairie grazing coincided with a marked change in climatic conditions in Miocene times. In North America, the cradle of horse evolution, there is evidence of general continental uplift and the

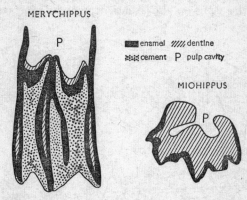

FIG. 43 The evolution of the horses – changes in the teeth. Vertical sections drawn to scale through upper molar teeth of an early browsing horse (*Miohippus*) with low-crowned teeth, and one of its descendants, an early grazing horse (*Merychippus*) with high-crowned teeth. Note the lengthening of the enamel ridges and the filling of cement. (*After Simpson*)

first widespread appearance of fossil grass seeds in the Miocene, which seem to indicate a general transition from hardwood forest to open prairie. It was in response to these environmental changes that the Miocene horses underwent such rapid and drastic modifications.

Although North America was the centre of the evolution of the horses, they spread into many other parts of the world at various times in the Cenozoic. These wanderings are illustrated in Fig. 40.

THE EVOLUTION OF LIFE

The living rhinoceroses are remnants of a group, now almost extinct, but formerly widespread and abundant. They arose from small slender ancestors in the Eocene. They have remained a group of browsers, and one of their Oligocene members, *Baluchitherium*, stood eighteen feet high at the shoulder, the largest land mammal ever to have lived (Fig. 44). One widespread contemporary of early man was the woolly rhinoceros, which lived in Europe during the last glaciation.

The living tapirs are also an ancient group, in many respects

FIG. 44 *Baluchitherium*, a hornless rhinoceros from the Middle Tertiary of Mongolia. It stood more than 16 feet high at the shoulder. (*After an exhibit in the American Museum of Natural History*)

remarkably similar in general structure to the ancestral odd-toed ungulates. Like the rhinoceroses, the tapirs underwent a great reduction in Pleistocene times, and they were formerly much more widespread. Their present restriction to Malaya and South and Central America is the result of this late Pleistocene extinction of geographically intermediate forms.

Two extinct groups of odd-toed mammals reached spectacular proportions. The titanotheres (Fig. 38) were massive, often horned creatures, the biggest of them eight feet in height at the shoulder, that were widespread in North America in Oligocene times. Their restricted browsing habits may have been a factor in their extinction. The more horse-like chalicotheres had clawed toes, which were an adaptation to digging out roots (Fig. 38). They were a restricted, but persistent group throughout Tertiary time, but, like many other large animals, they disappeared during the Pleistocene glaciation.

Of these five groups of fossil odd-toed ungulates, the tapirs, rhinoceroses, and horses (in their wild state) survive as precarious remnants. The odd-toed ungulates have, in spite of their past greatness, passed their evolutionary peak, and are now in steady decline.

They have been largely replaced by the now dominant even-toed artiodactyls. Their present diversity and abundance is a relatively recent development, although they have evolved from an Eocene stock. The pigs and their relatives are the most primitive living members of the group. Some of their Tertiary forebears were as large as horses. The artiodactyls are adapted to an astonishing variety of environments – such as the amphibious hippopotamus of tropical Africa, camels and llamas of deserts, deer and elk of forests and tundra regions, giraffes of the savanna, wild sheep and goats of the mountains, and cattle and bison of the plains. Many of these creatures were formerly far more widespread: even during the Pleistocene there were hippos in England, while throughout most of their history the camels were confined to North America. The rapid expansion of the artiodactyls is regarded by many as the direct result of the cud-chewing habits of many of their members. This may or may not be so, but certainly the ability to 'eat in haste and digest at leisure' confers some advantage in the savage world of carnivorous competitors.

There are two other groups of ungulates which have been important throughout most of Cenozoic time. One group (p. 246) was peculiar to South America. The other group, the elephants, needs little introduction, although many of their ancestors have a thoroughly unfamiliar look. They appear to share a common ancestor with the African conies and the sea cows. The earliest 'elephants' were pig-like animals, found in the Eocene of Egypt.

Like so many other larger mammals, elephants were widespread until comparatively recent times. The woolly mammoth, for example, was abundant throughout North America, Africa, Europe, and Siberia, during the last glaciation. It provided a common source of prey for early man, and is frequently depicted in cave paintings (Fig. 50J).

Other groups, though less conspicuous than these large mammals, played a prominent part in the economy of Cenozoic life. The rodents, for example, have been widespread and varied since the early Tertiary. They are a very successful group, widely adapted to the variety of habitats and modes of life, and represented by rats, mice, beavers, porcupines, and squirrels. Like their relations, the rabbits and hares, many of them are also enormously abundant as individuals.

SOUTH AMERICAN MAMMALS

The mammals of which we have spoken are characteristic of North America, Europe, Asia, and Africa, and in general these areas today still exhibit a strong faunal similarity (although we have noted some differences). The explanation lies in their geological history, for they were periodically though frequently connected during Tertiary time. There were, of course, and still are local barriers and restricted connexions and therefore differences; but there remains a broad uniformity between all these continental faunas. Once we look at the Cenozoic faunas of South America, however, we find a completely distinct group of animals. South and Central America (the Neotropical Realm, as zoologists call it) was linked to North America in early Cenozoic times. From then until the late Pliocene, however, 60 million years later, it was isolated. It is this isolation which is the clue to its distinctive fauna, just as it is in the case of Australia. Both the marsupials and the placental mammals were established at the dawn of the Cenozoic era, and we have seen that in the Northern Hemisphere the marsupials were

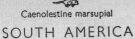
Caenolestine marsupial

SOUTH AMERICA

Shrew

NORTH AMERICA

Marsupial carnivore

Wolf

Camel-like litoptern

Camel

Horse-like litoptern

Horse

Toxodont

Rhinoceros

Homalodothere

Chalicothere

FIG. 45 Convergent evolution among North and South American mammals. Drawn to same scale. (*After Simpson*)

completely overshadowed by the placental expansion. But in the Southern Hemisphere, they held their own. They became, in fact, the chief carnivores, some of them being wolf-like creatures and others growing to savage predators, the size of lions, and bearing a remarkable resemblance to the sabre-toothed cats of other areas. These two latter groups were quite unrelated in origin and their similarity was due to similar coincidental adaptation to similar modes of life – a perfect example of convergent evolution (Fig. 45).

The hoofed marsupial mammals of South America were an equally distinctive group, including five orders of great variety. Some of these creatures were of considerable size: *Toxodon*, a massive Pleistocene form, stood six feet high at the shoulder, but many were dog-sized and others still smaller. Many of these South American forms were strikingly similar to contemporary hippopotami, horses, camels, llamas, elephants, and other creatures in other parts of the world, but here again the resemblance was the result of adaptation to similar environments, rather than immediate genetic affinity. These hoofed mammals flourished throughout the Tertiary, but, like the marsupial carnivores, they became extinct in Late Pliocene times, as hordes of North American placental mammals filtered southwards over the newly raised Panamanian land bridge between the two continents. It was not a one-way traffic, but the South American mammals had almost no success in their colonization in the opposite direction. They represent a unique and major group of hoofed mammals, and their multiplication stemmed from their isolation. But just as, in one sense, isolation may be productive, so, in another, its end may bring catastrophe, even after some 60 million years of progress. For the isolation of the creatures had brought only 'partial' natural selection; with the arrival of North American carnivores and ungulates came competition for which even the spectacular diversity of an era's cloistered change was no match.

The other distinctive group of South American mammals are the edentates, a group that includes two rather different types of animal (Plate 8). One group includes the living arma-

THE AGE OF MAMMALS

dillos and the extinct glyptodonts. These were creatures some eight or nine feet in length, which were protected by a huge domed armour plate. In some of them the tail ended in a savage club-like structure. The second group includes the herbivorous sloths and the anteaters. In late Cenozoic times some ground sloths achieved spectacular proportions, one monster (*Megatherium*) reaching a length of twenty feet – rather larger than a good-sized elephant. Both these creatures and the glyptodonts spread into North America when the land connexion was re-established between the two continents. Their remains have been found associated with stone weapons of early man and they probably became extinct in only comparatively recent times. The other members of the South American fauna, the New World monkeys and some rodents, were later arrivals which migrated into the continent in late Eocene times. They are both characteristic of the continent, but their relatively closer resemblance to forms from other continents is a measure of their shorter period of isolation than other South American forms. More recent arrivals, deer for example, are even more similar to their North American counterparts, while a few other forms, such as tapirs and llamas, are remnants of a group now extinct in North America.

This distinctive South American fauna of glyptodonts, giant sloths, carnivorous marsupials, and diverse ungulates illustrates the overriding importance of geographical isolation in the evolutionary process. The same is true of Australia, which appears to have been isolated since Cretaceous times. There, as we have already seen, the marsupials faced no competition from the placentals and their diversity in that continent seems to be the direct result. Isolation is a factor that operates at all levels – from the level of whole continental faunas, as in this case, to the level of a single species, as in many birds. For wherever populations, of whatever size, are cut off from interbreeding with their neighbours, there are new species in the making.

THE EVOLUTION OF LIFE

THE RETURN TO THE SEAS

As in amphibia, so in reptiles: as in reptiles, so in birds: as in birds so in mammals, the mastery of a new environment and the adaptation that went with it brought a paradoxical return to the old. And the mammalian return to the sea, like that of the reptiles and the birds before it, was a marked success. What a striking example this is of the endless diversity and potentiality of the form of living things! For here are mammals, the most fully and perfectly adapted of all animals to life on the land, and at the very beginning of their radiation we find them back in the sea – equally well-adapted.

Three quite distinct mammalian groups have made this return. The whales and dolphins (the cetaceans) represent the epitome of adaptation to marine life. Fishlike in form and general structure – propelled by the powerful tail, balanced by the well-formed fins – yet still essentially mammals in their (modified) lungs and in the suckling of their young. The oldest fossil whales come from the Eocene, but even at that early time they are fully differentiated and provide no real clue to their ancestory. It seems probable that they arose from some form of carnivorous creodont ancestors.

No other mammals approach the whales in the perfection of their adaptation to life in the seas. Almost every typical mammalian character is beautifully modified in the whales for their new existence, and those, such as the hind limbs and the hair, which have not been bent to a new function have been lost. In size, as well as form, these noble creatures are remarkable: they are the largest animals ever to have lived. The blue whale is 120 feet long and weighs 150 tons!

The dolphins are no less successful, and in them there is a compelling parallel to the ichthyosaurs. Fish, ichthyosaurs, and dolphins show an incredible resemblance to one another. It was in fact the empty seas of the extinct ichthyosaurs and plesiosaurs that the cetaceans colonized, and their resemblance is far from coincidental. It is another example of convergent

THE AGE OF MAMMALS

evolution – that marvellous testimony to the truth of natural selection.

The sea cows (sirenia) include the dugongs of the Indian Ocean and the manatees of the tropical Atlantic and represent another group of mammals which have become highly adapted to life in the seas. Their closest living relatives are the elephants and the African conies, but they have a long independent history of their own, for they were already well established by Eocene times. They are ungainly creatures, known as Sirenia, and it seems paradoxical that such thoroughgoing seafarers should be most closely related on the one hand to the tiny conies and the other to the elephants. Yet, for all their strange appearance, the body structure points clearly to a common origin for the three groups.

The seals, sea lions, and walruses are the third mammalian group to have returned to the seas, though their adaptation is less complete than that of the whales and sea cows and they are still partly (though rather helplessly) terrestrial in their habits. They are predacious carnivores, sharing a common ancestor with bears and cats, and probably arose from some dog-like stock in Oligocene times. They first appear as fossils in the Miocene.

We have now seen something of the great mammalian radiation that took place in Cenozoic times. It carried mammals into every niche and corner of the earth, into the air, and into the seas. But there is one other line of mammalian evolution that is of particular interest, for we are its children.

Chapter 14

THE ADVENT OF MAN AND HIS KIND

We have covered more than thirty orders of mammals in two brief chapters and to devote a whole chapter to the one remaining order may seem excessive. But the Order Primates includes lemurs, tarsiers, monkeys, apes, and man himself (Fig. 46), and if it is true that man's chief attribute is his self-consciousness, then our preoccupation with our own kin may at least be regarded as human frailty. The primates are essentially an arboreal group and are therefore, like the birds, rather poorly represented as fossils. Because of their arboreal habits they have tended to remain rather unspecialized animals, especially in their limbs and teeth, and they have developed complex brains and acute vision, which attributes have been of major importance in their subsequent development.

The oldest primates are lemur-like creatures which occur in the Paleocene, and which are probably the descendants of some arboreal insectivores, not unlike the living tree shrews. The living lemurs, aye-ayes, and bush babies are archaic remnants of this early group. These small creatures, looking rather like a combination of squirrel, shrew, racoon, and monkey, are superficially quite unlike the higher apes, but the two groups are linked by the tarsioids – tiny, wide-eyed, arboreal, jumping, rat-like creatures from the East Indies. These animals are intermediate in character between living lemurs and monkeys, and it seems very probable that their Eocene forebears were the transitional forms through which the monkeys arose from the primitive lemurs (Fig. 46).

MONKEYS, APES, AND MEN

Monkeys, apes, and men constitute the third and most advanced primate group. The marmosets, squirrel, spider, and

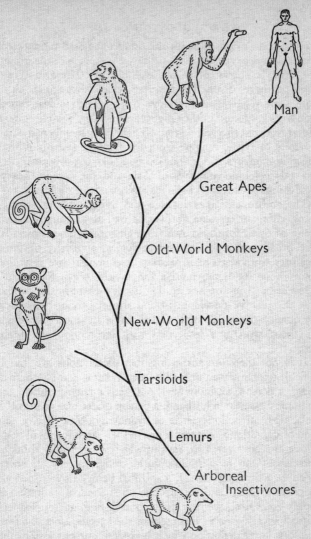

FIG. 46 The family tree of the primates.
Diagrammatic representation of the main groups of primates. Not to scale. (*After Romer*)

the capuchin monkeys, with the other flat-nosed monkeys of South America, are the more primitive members of the group, but little is known of their fossil history. The Old-World monkeys have distinctive features which distinguish them from their relatives of the New World. Their oldest representatives are found in the Oligocene, and they have become a widespread and successful group. Their living members include a variety of types, ranging from the familiar and ubiquitous malaque monkeys to the long-limbed, slender, arboreal, herbivorous colobine monkeys of Asia and the savage, heavy, carnivorous, quadruped baboons and mandrills of Africa.

The anthropoid apes include the living gibbons, orangutans, chimpanzees, and gorillas, as well as an equally varied group of fossil forms, known from Oligocene times onwards. The broad anatomical and physiological similarity of these creatures to modern man needs no elaboration: we resemble them 'bone for bone, muscle for muscle, organ for organ'. There is no structure in man that is not present in the gorilla: only the relative development and size of certain body parts are quantitatively different, and of these the most important are those structures associated with locomotion and brain size.

It is not therefore surprising that palaeontologists find it difficult to define man in fossil terms. What is man? 'A human being, a person, an individual, one with manly qualities' the dictionary asserts: 'a husband, a person under one's control' it adds, with an oblique sense of synonymy. Now clearly this kind of definition is not much use in the search for fossil man. There are, however, two ways in which we generally think of ourselves. One is a philosophical, religious, moral, social sense – 'a thinking animal, a spiritual being, a self-conscious moral individual' and so on. The other kind of definition is purely zoological, we may define ourselves in anatomical and physiological terms as members of the species *Homo sapiens*. It is this latter definition which best suits our purpose in the present discussion, not because it is more fundamental or more important than the first, but simply because it is the only one

THE ADVENT OF MAN AND HIS KIND

we can apply to fossil man. This implies neither denial, doubt, nor disrespect towards the first type of definition: indeed the two definitions are not discrete, but are intimate parts of a single entity – man. To deny man's uniqueness on the one hand is just as naïve as it is on the other hand to deny that this uniqueness is the result of his bodily structure, and particularly his brain. It was Charles Darwin who first demonstrated the descent of man: and it was also Darwin who spoke of man as the wonder and glory of the Universe.

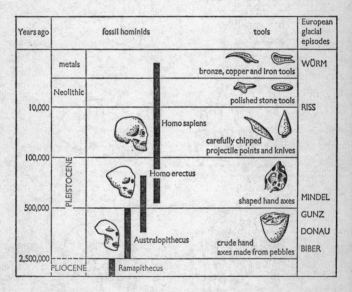

FIG. 47 Table showing the division of Pleistocene time and the relative durations of human cultures. The periods of existence of various genera are plotted on the right. (*After Ebert, Loewy, Miller and Schneiderman*)

(Not drawn to scale)

The path by which *Homo sapiens* has arisen is now fairly well

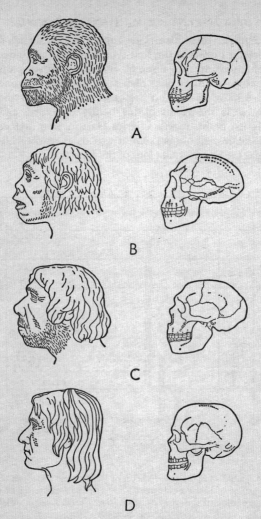

FIG. 48 Fossil man. A group of skulls, oldest at the top, showing changes in the skull of fossil man.
A *Australopithecus*, B *Pithecanthropus*, C Neanderthal man, D Cro-Magnon man. (*Partly after Simpson, Pittendrigh, and Tiffany*)

known, although, as we should expect, the record is far from complete. We can best illustrate this evolutionary sequence by describing three groups of fossil man-like creatures, all of which agree very closely with the probable human ancestors predicted by Darwin and others long before actual fossil remains were known. In fact the only fossil 'man' that did not fit into the general evolutionary sequence was *Eoanthropus dawsoni*, Piltdown man, and this one-time discrepancy has now been resolved (*see* p. 53).

ANCIENT MAN

The oldest true hominids are found in the Late Miocene and Pliocene rocks of India, Europe and Africa. They belong to a still imperfectly known genus, *Ramapithecus*. These were creatures about three or four feet tall, although it is not known if they stood erect. They had a semi-circular arrangement of teeth that is far more similar to man than that of apes, and suggests a generalized diet. The specimens from Africa have been dated at 14 million years old. They seem to have inhabited warm forest environments.

Australopithecus ('southern ape') was the name given by Professor Raymond Dart in 1925 to a skull which he found in cave deposits in Bechuanaland (Fig. 48A). Similar skulls, and other associated bones, were subsequently found by Dr Robert Broom near Johannesburg, and their occurrence suggests that they are Lower Pleistocene or Upper Pliocene in age – approaching two million years old in round figures, although other occurrences suggest they may have existed for a total span of some four million years. *Australopithecus* was a pygmy-sized individual, with a curious mixture of human and ape-like features. The skull, for example, had massive protruding chimpanzee-like jaws, and a brain capacity less than half that of modern man, although rather larger than a living chimpanzee. But in other respects it was man-like: it had reduced brow ridges, a rounded rather than ridged back of the skull, a rather human forehead, and teeth which show a semi-circular arrangement, in contrast to the more angular and straight-tooth arrangement of the apes. In their semi-erect posture, too, these creatures were man-like and their remains are widespread in the Transvaal. Other specimens are known from the Near East and China. They seem, in fact, to have been well adapted for life on the semi-arid plains, in contrast to the characteristic tropical forest existence of the apes. It is not yet clear whether they were cave dwellers, or whether their remains were carried to such caves by carnivores which preyed upon them. There is no clear evidence of any power of speech, but it has been suggested that australopithecines used stone tools in East Africa and bone tools in South Africa. A number of baboon skulls have been found near the original site, however, which have depressed fractures, and it is possible that these may have been inflicted by the forerunner of the more modern 'blunt instrument'. If these baboons were preyed upon by *Australopithecus*, it would certainly imply a considerable fleetness of foot and skill in hunting.

What then is *Australopithecus* – a coincidental man-like ape, or a transitional form between apes and man? It seems improbable that the large number of very close structural similarities to the hominids could occur independently in another group, and the similarity strongly suggests that the australopithecines are more closely related to 'men' than to apes. Whether or not they are direct human ancestors is another matter. The fossils that are known suggest the existence of considerable variation, perhaps representing several species including a lithe carnivorous type and a more robust group. It seems very probable that the carnivorous group (*Australopithecus africanus*) represent the ancestors of man.

The immediate ancestors of *Australopithecus* are not yet known, but they may well have resembled *Ramapithecus*, which shows characteristics intermediate between australopithecines and a great variety of anthropoid apes which lived in Europe, Asia and Africa during Miocene times. It is amongst these that our remotest ancestors will probably be found. Many palaeontologists believe that the dryopithecines, a group of very

THE ADVENT OF MAN AND HIS KIND

generalized apes, which were widespread in the Old World in Miocene and Pliocene times, may be the long-sought relatives.

It is a matter of definition as to whether we call the australopithecines 'human' or 'men'. We have shown that for all their ape-like appearance they belong in the hominid family to which man assigns himself. Most palaeontologists, however, regard man as a tool-making creature and limit the title to later hominids which had developed the intelligence to manufacture such tools. It is these creatures to which we must now turn.

The history of '*Pithecanthropus*' (the 'ape man') goes back to a collecting expedition made by a young Dutch army physician, Eugene Dubois. Dubois was fascinated by the problem of human origins and he reasoned that the presence of the living orangutan in the Indo-Malayan area suggested that it was there that man's earliest ancestors might be found. Between 1888 and 1895 he undertook an extensive search in Central Java, and in 1891 discovered a skull and later a thigh-bone near the hamlet of Trinil, on the Solo River. The remains aroused the most violent controversy, for although the low, heavy, rather small skull with massive brow ridges was clearly ape-like in form, the thigh-bone left no doubt that the creature stood erect.

Subsequent discoveries in Java, Africa, Europe, and near Peking in China have provided a substantial number of specimens of '*Pithecanthropus*', on which a dependable reconstruction may be made. In some respects '*Pithecanthropus*' was certainly ape-like (Fig. 48B). It was a rather squat creature, about five feet high, with heavy beetling brows, a retreating forehead, and powerful jaws. But it also stood erect and both the nature and the arrangement of the teeth are man-like, although its brain capacity of about 1,000 c.c. is greater than that of the largest living apes (about 500 c.c.) but less than that of modern man (about 1,350 c.c.). These man-like characters are so striking that most anthropologists now assign *Pithecanthropus* to our own genus, *Homo*, in a separate species, *H. erectus*. Actually the Chinese and Javanese specimens show some differences and were once classified as different genera. It now seems clear, however, that they represent a single species, *H. erectus*, that was rather variable in form.

We now face the same problem that we faced in the case of the australopithecines – 'man or beast?'. If we had only the skeleton of *Homo erectus* on which to form judgement, we should be equally uncertain. But associated with *Homo erectus* are implements of quartz and shaped bone tools, as well as the charred bones of deer. Some skulls of *Homo erectus* are broken in such a way as to suggest that the brain was extracted from them. Clearly, therefore, *Homo erectus*, for all his cannibalistic tendencies, was a skilled hunter and craftsman, already familiar with the use of fire, and capable of the conceptual thought required to design and make an implement for some future, but not immediately apparent, need.

Homo erectus, then, represents the earliest man – savage, cannibalistic, of low intelligence, ape-like in aspect – but, for all that, a man, broadly comparable in his level of cultural development, as represented by weapons and tools, to some of the races of mankind which still survive. He (as we may now legitimately refer to him) existed during Early and Middle Pleistocene time, about 500,000 years ago, and it seems very probable that *Homo erectus* is ancestral to our own species, *Homo sapiens*.

We have already seen that hominids are rare as fossils, and our discussion of the rather scanty remains of both *Australopithecus* and *Pithecanthropus* has emphasized the point. The remains of more recent men are rather more common, though never abundant, but once man became a tool maker a second type of human fossil became available, and we have seen in the case of *Homo erectus* how much these implements can reveal of early man. These implements are much more resistant to erosion than are bones, and they are relatively common and widespread throughout Pleistocene deposits (Fig. 49).

The Pleistocene began about two million years ago (Fig. 47) and we have seen that it was represented by a succession of glacial and interglacial episodes in the Northern Hemisphere and pluvial and interpluvial episodes in the Southern Hemi-

THE ADVENT OF MAN AND HIS KIND

sphere. It was within this frozen world that early man lived, the companion of the mammoth, reindeer, and woolly rhinoceros during the harsh glacial periods, and the hippopotamus and straight tusked elephant during the milder interglacial episodes.

The oldest tools were used but not made. Tool-using man seems simply to have picked up whatever conveniently shaped stones or flints came to hand. It was only subsequently that crude chipping was used as a method of improving the usefulness of these flakes, and the variety of tool and the skill and degree of workmanship show a steady increase throughout Pleistocene time. This Palaeolithic period of culture was marked by the use of chipping only, but about 7,000 years ago the practice of grinding and polishing stone tools was introduced, and this later period is known as the Neolithic. It is therefore possible to obtain a sequence of tools and artifacts which may be used as a chronological scale in the study of ancient man.

We should not expect, of course, that isolated groups of men would everywhere make exactly the same type of tools at the same period of time, for the 'Stone Age' still persists among some isolated tribes today. Allowances have therefore to be made for these local industries, and indeed it has recently become possible to trace the distribution of certain types of tool and recognize ancient dispersal centres and trade routes. There was, for example, an ancient factory just south of Corndon Hill in Montgomeryshire, where a tough intrusive rock (picrite) was worked for axe-hammers.

With this graduated succession of tools may be correlated other aspects of man's development. Certain skeletons, for example, are associated with particular tools; ceremonial burial was first practised in Mousterian times, and so on.

What were they like, these men of the Old Stone age? All of them are assigned to our own genus *Homo*, and this is a measure of their resemblance to modern man. Most of their remains have been found in Europe, and the most ancient of them is the single lower jaw of Heidelberg Man (*Homo heidelbergensis*), which some regard as a racial variety of *Homo erectus*, discovered in 1907 near Heidelberg. No tools were found with it, but associated fossil animals suggest that he existed during the first interglacial period, about 500,000 years ago. The powerful massive jaw shows that Heidelberg Man was a chinless creature, possibly with a rather heavy face, although the teeth are characteristically human.

NEANDERTHAL MAN

The first remains of Neanderthal Man were discovered over a century ago near Düsseldorf and consisted of a skull cap and limb fragments. Since then more complete fossils of the same type have been found, representing a race similar to, but apparently distinct from, modern man. Although once regarded as a distinct species, *Homo neanderthalensis*, Neanderthal man seems more likely to have been only a race of our own species, which was widespread throughout Europe, Siberia, and North Africa during the last glacial period. Other fossil remains of comparable age from Croatia, Rhodesia, and Java which show a close resemblance to these typical Neanderthalers may indicate an even wider distribution. These men had very massive brow ridges, a retreating forehead, and a heavy and prominent but chinless jaw (Fig. 48c). This was emphasized by their short, stooped, thick-set, muscular body, although their average brain size was rather larger than that of modern man.

Neanderthal Man was formerly regarded as the ancestor of *Homo sapiens*, but it is now clear that he was both too specialized in structure and too late in time to have occupied such a position. He seems rather to be a distinct but parallel branch of the same genus, sharing a common ancestor with modern man.

We know a good deal about the way our Neanderthal relatives lived their lives during the last harsh days of the Ice Age. They made a variety of tools including hand axes, scrapers, points, knives, chopping blocks, and possibly bolas, carefully flaked and chipped in characteristic style (the Mousterian culture). The oldest representatives of the race appeared during the

THE ADVENT OF MAN AND HIS KIND

equable conditions of the third interglacial period, but their descendants lived in the bleak tundra wastes that spread across Europe along the margins of the continental ice sheets of the fourth and final glacial advance. Many of the Neanderthalers were cave-dwellers, and frequently used fire. They hunted a variety of animals, including savage cave bears, mammoths, and rhinoceros, which they probably hunted with wooden spears. These weapons themselves are unknown, but evidence of their use is preserved in a Neanderthal-type skeleton from Mount Carmel, in which a spear had been driven through the thigh into the pelvis. This sombre record of ancient conflict is not altogether unexpected, for at least some of these ancient men were probably cannibals.

It was with the coming of Neanderthal Man that we catch a first glimpse of man's inmost thoughts. In some caves the skulls of cave bears were carefully arranged and stored, perhaps providing a shrine or talismans for the hunt. The dead too were often carefully buried in a foetal position with weapons, in such a way as to leave little doubt that Neanderthal man had some hope or belief of survival beyond death.

These people lived in Europe for 100,000 years and then, about 30,000 or so years ago, they became extinct. Their disappearance is not easy to explain: perhaps it was the result of the increasing rigours of life in that frozen world of the Pleistocene, perhaps it was the result of increasing competition from our own species. Some have suggested that the disappearance was the direct result of the invasion of Europe by a new population of *Homo sapiens*, who came from the East. We do not know. But with the passing of Neanderthal man, this new race spread across the face of Europe, and the supremacy of *Homo sapiens* was established.

MODERN MAN

Two fragments from the gravels of the Thames at Swanscombe in Kent are associated with Acheulian implements which almost certainly represent remains of the second interglacial

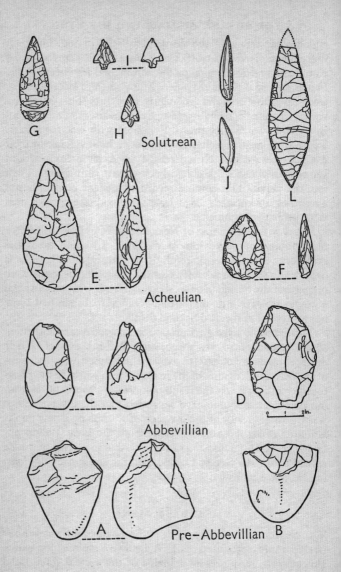

THE ADVENT OF MAN AND HIS KIND

period. Although these bones are relatively thick they show a very strong similarity to those of modern man, and indicate that *Homo sapiens* may have existed for at least 200,000 years. Another more complete skull from Steinheim, near Stuttgart, which may be of comparable age, has a smaller brain and prominent brow ridges which are reminiscent of *Homo erectus*. In spite of this it, too, is generally regarded as a primitive member of our own species, although its resemblance to *Homo erectus* seems a clear pointer to its ancestry.

Apart from these two glimpses, there are no records of our own species until the mass replacement of the Neanderthalers by the Cro-Magnon race. To our anthropomorphic senses there is a profound contrast between the two peoples: Neanderthal man, with heavy browed, low, sloping head, short and brutish, and Cro-Magnon man, finely built, tall and muscular, with large brain and handsome facial features, indistinguishable from those of modern man (Fig. 48D). He competed with Neanderthal man in the frozen wastes and tundra, but he competed on unequal terms. The tools of Cro-Magnon man are far superior to those of Neanderthal: they are more finely worked, smaller, more specialized; they are fashioned from ivory and bone as well as stone, and later the first arrowheads appear.

In other ways too the culture of Cro-Magnon man was distinctive. He not only buried his dead, he buried them with reverence and ceremony and ornament. One tomb of twenty

---

FIG. 49 The development of Palaeolithic cultures (see Fig. 47). Note the increasing use of chipping and gradual refinement of workmanship in fashioning the various tools.

A–B Pre-Abbevillian tools, pebble tools from Tanganyika; C–D Abbevillian tools: C hand axe, 150-feet terrace of River Thames, Berks, D hand axe, derived, Chelles-sur-Marne, France; E–F Acheulian tools: E lava hand axe, Kenya, F ovate hand axe, near Amiens, France; G–L Solutrean tools: G spear point, Stillbay, S. Africa, H arrowhead, Parpallo, Spain, I arrowhead, Morocco, J knife point, Chatelperron, France, K knife point, Laussch, France, L 'laurel leaf' blade, Solutre, France. (*After Oakley*)

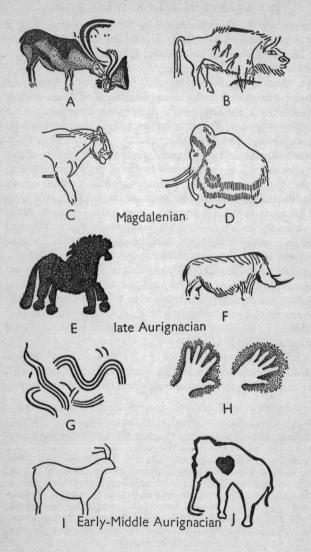

THE ADVENT OF MAN AND HIS KIND

individuals is known in which the bodies are flanked with carefully arranged mammoth bones. The skeletons of the dead were often adorned with necklaces of shells and fossils and coloured with red ochre. And with this reverence for death went a zest for life. Cave paintings of great beauty and sensitivity are known from Russia to the Pyrenees, with exquisitely executed polychrome paintings and decorative carvings on bone, and carefully and beautifully carved figurines are common. With the coming of these things, we reach a great landmark in our history, for here is the end of the evolution of man the animal, and the dawn of man the unique – a self-conscious, rationalizing, artistic, worshipping creature. All these capacities we can trace in Cro-Magnon man.

This was not of course the end of man's evolution: the Cro-Magnon is only one of many races of our own species and even within Cro-Magnon peoples we can detect some racial differences.

Chipped and flaked tools were gradually supplanted by the ground and polished artifacts of Neolithic man, and these in turn by tools of bronze and later of iron. Language, tradition, the growth of communities and social structures, and increasing control of his environment, all these have played a major part in man's subsequent development. They are the stuff of anthropology and history, not of palaeontology. And yet there is an intimate bond between the two, for all these things are the products of one organ, whose growth we have traced from

---

FIG. 50 The development of Upper Palaeolithic cave art (see Fig. 48), A–D Magdalenian: A 'reindeer fighting', male in black, female in red. Font-de-Gaume, France, B engraving of wounded bison, Niaux, France, C engraving of cave lion, Combarelles, France, D engraving of woolly mammoth, Font-de-Gaume, France; E–F Late Aurignacian: E horse in black manganese oxide, Lascaux, France, F woolly rhinoceros in red-ochre, Font-de-Gaume, France; G–J Early-Middle Aurignacian: G snake-like scribbles in yellow-ochre, La Pileta, Spain, H hands outlined in red-ochre, Castillo, Spain, I engraving of deer, Pair-non-Pair, J elephant (with heart) in red-ochre, Pindal. (*After Oakley*)

the earliest primates through to man. In no other group has there been such an enormous increase in brain size and complexity. It is this which is the centre of the new evolution, for with the coming of modern man the established evolutionary process has entered a new phase. For the first time in the long history of life, a species has developed with the power to control the future course of evolution.

Chapter 15

THE TEEMING SEAS

ANY history of life is apt to become lopsided – to begin in the seas, but then to ignore them and to concentrate on the life of the land. This is understandable. We ourselves live on the land and our terrestrial neighbours are far more familiar to us than the creatures of the seas. Even if we live near the coast, we usually become familiar with only the handful of animals and plants of the shore. Yet the seas cover almost three quarters of the globe and support an abundance of living things far greater than those of the land.

The seas have an average depth of 12,000 feet and, although the concentration of living creatures is greatest within the upper zones, there are no lifeless areas or zones in this enormous body of water. Even the ocean depths at 35,000 feet (six miles) support a variety of organisms, living in total darkness, at about $0°$ C., under enormous pressure.

The animals and plants of the seas exist in countless numbers. Most groups of animals and plants arose within its waters. All the animal phyla are still represented there, and about one third of all the classes are confined to the seas. Only in numbers of species are the oceans less prolific than the land, and the reason for this presumably lies in the relatively greater environmental uniformity and the much lower degree of isolation which exist in the seas. But the uniformity and continuity is not so great that there are no differences: the fauna of the tidal zone is quite distinct from that of the continental shelf, for example, and many marine animals undertake seasonal migrations of thousands of miles.

It is easy to overlook this inexhaustible richness of the seas, but it forms a major part of the history of life, of no less importance than the history of life on the land.

The living algae and diatoms are the grass of the oceans and

THE EVOLUTION OF LIFE

they occupied a similar position throughout Mesozoic and Cenozoic time. The myriads of foraminifera that lived throughout the Mesozoic and Cenozoic were very different from the majority of their Palaeozoic ancestors. Although individual genera have come and gone, there is a broad similarity between these fossil forms and those that still exist today, and the same is true of the radiolarians.

The sponges of both the Mesozoic and the Cenozoic were not greatly different from those of our present seas. They were a widespread but unspectacular group, the calcareous forms being generally characteristic of the more shallow waters, and the siliceous of the deeper seas in which limestones were deposited. Their tiny fossil spicules are abundant in some of the flints of the Chalk, and they may have been important contributors to the formation of this siliceous rock.

Mesozoic corals differed chiefly from those of the Palaeozoic in their sixfold, rather than fourfold, symmetry. Many were single forms, but others were colonial reef builders. Such reefs were widespread in the past, and in the Jurassic, for example, existed as far north as Scotland, at a great distance beyond their present restricted limits. This implies either a much greater tolerance for extreme climatic conditions in the past, or, more probably, a northward extension of tropical waters.

Mesozoic bryozoans were quite different in structure from those of the Palaeozoic, and they underwent a great expansion during Cretaceous times. They remain a widespread, though

---

FIG. 51 Mesozoic invertebrates.
A–D Cretaceous; A *Monopleura*, a rudistid pelecypod, length 2 inches, B *Pecten*, a clam, width 1½ inches, C *Turritella*, a gastropod, length 2 inches, D *Inoceramus*, a pelecypod, some species measured 4 feet across the shell; E–I Jurassic: E *Gryphaea*, two species, a pelecypod, length 2 inches, F *Pentacrinus*, a stalked crinoid, 'armspan' about 3 inches, G *Goniothyris*, a terebratulid brachiopod, length 1½ inches, H *Hemicidaris*, a regular echinoid, side and upper views, diameter about 1 inch, I *Spiriferina*, the last of the spiriferid brachiopods, width 1 inch. (*Partly after Dunbar*)

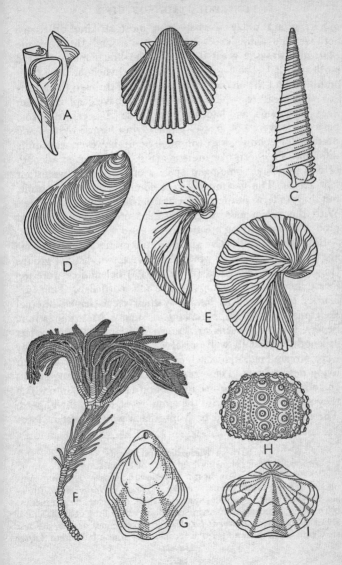

271

THE EVOLUTION OF LIFE

minor, group today. Perhaps the most striking difference between the animals of Palaeozoic seas and those of the Mesozoic was the great decline of the brachiopods. Of the hordes of Palaeozoic groups, only the rhynchonellids, terebratulids, and the eternal *Lingula* persisted through the Mesozoic. The creatures which seem largely to have displaced them were pelecypods, and later the gastropods (Fig. 51). Mesozoic pelecypods were not greatly unlike living forms; oysters and pectens, for example, were common. Some, however, were quite different. The horn-like rudistids (Fig. 51A) reached a length of two feet, and some giant species of *Inoceramus* were four feet across. The gastropods were common in the Mesozoic, but they underwent their greatest expansion in Cenozoic times. With the pelecypods, they represent one of the dominant groups amongst living creatures of the sea.

But the most distinctive marine intertebrates of the Mesozoic are the cephalopods, of which there were two distinct groups (Fig. 52). The two-gilled (dibranchiate) belemnites were the forerunners of the living squids and cuttlefish. They are usually represented as fossils by their cigar-shaped internal skeletons, which are abundant in many Mesozoic strata. Several hundred species are known, the largest being six feet in length. Unusually well-preserved specimens show that they were streamlined, tentacled creatures, very similar to their living descendants. These cephalopods, like the living squids and octopus, rely for protection, not on a heavy shell into which they may withdraw, as did the ammonites, but on their speed. The mantle is modified both into flaps which assist in loco-

FIG. 52 Mesozoic cephalopods.

A *Ceratites*, Triassic, showing beginning of crenulation in the suture line, diameter about $2\frac{1}{2}$ inches; B *Phylloceras*, Jurassic, showing complexity of suture line, diameter about 2 inches; C *Normannites*, Jurassic, showing complex ornamentation of shell, diameter about 2 inches; D *Turrilites*, Cretaceous, showing complexity of coiling of shell, length about 5 inches; E *Belemnitella*, Upper Cretaceous, one of the last survivors of the belemnites; F a reconstruction of a belemnite. (*After Hayes*)

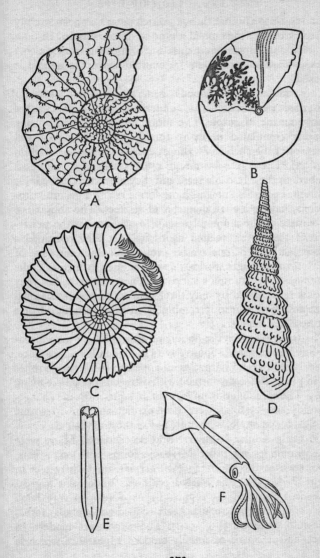

motion and into a funnel, through which water may be so forcibly ejected as to give the animal a type of jet propulsion. In both living and fossil forms an ink sac is also present, from which a cloud of dark material may be emitted when the animal is attacked.

The ammonite cephalopods, generally regarded as being four-gilled (tetrabranchiate), underwent profound changes throughout the Mesozoic. The suture line of the shell became intensely crenulated in many forms, while the shape and ornament of the shell show almost every conceivable variation (Fig. 52). These free-swimming creatures lived in countless numbers in the Mesozoic seas, and they are amongst the most useful index fossils known. Some forms reached gigantic proportions, having a shell diameter of six feet. The abundance and variety of the ammonites tend to overshadow the persistence of the closely related nautiloids, yet they remained a widespread though conservative group, and when the close of the Mesozoic brought world-wide and sudden extinction to the ammonites, the nautiloids survived. They still exist, but they are now represented by only three genera, which are confined to tropical waters: exotic reminders in a modern world of a greatness that is past.

The trilobites were the dominant arthropods of the Palaeozoic but Mesozoic arthropods have a more familiar appearance. Lobsters, crabs, and barnacles all appeared in Mesozoic times and in places the seas swarmed with ostracods not unlike living forms. The echinoderms underwent an equally drastic change. Blastoids and cystoids were now extinct and the Mesozoic crinoids, for example, were quite unlike those of the Palaeozoic, but broadly similar to their living descendants. Many were free-swimming, but other Mesozoic forms had long stems, some, such as the surviving *Pentracrinus*, up to fifty feet in length (Fig. 51F). The related echinoids underwent a great expansion. Some of the regular forms developed very large spines and a host of irregular heart-shaped and globular forms came into existence (Fig. 51H). These too were modern in aspect, and it is this essentially modern appearance which is

THE TEEMING SEAS

most characteristic of Mesozoic and still more of Cenozoic faunas. All the main groups of living marine invertebrates were well established by Mesozoic times, and apart from the extinction of the ammonites and belemnites the only changes which took place were of a comparatively minor kind.

The vertebrate dwellers of the seas have undergone much more change. The rise and extinction of the great marine reptiles, the evolution of sea birds and ocean-going mammals are all events which have taken place within the span of Mesozoic and Cenozoic time. The fish, too, have undergone considerable change.

The Mesozoic history of the sharks was one of increasing specialization and modernization. The primitive jaw suspension was replaced in most forms by the more advanced and efficient attachment of modern fish, and many Mesozoic sharks were similar in their general streamlined body and strong fins and tail to living forms of the open seas. Others became bottom dwellers, with flattened bodies, whip-like tails, and teeth adapted to grind the shell fish on which they feed. These were the ancestors of the skates and rays and they first appeared in early Mesozoic times.

But in spite of the persistence and the success of sharks, it is the bony fish that represent the peak of vertebrate adaptation to an aquatic life, not only in the seas but also in inland waters. Their Palaeozoic ancestors included, as we have seen, the little ray-finned palaeoniscids of the Devonian, as well as the lobe-finned coelacanths and lungfish. Although the living sturgeon and a few other fish are remnants of early ray-finned fish, most of the primitive forms were replaced in Mesozoic times by the holostean fishes (Fig. 26).

These holosteans, of which the North American garpike and bowfish are the only survivors, had a more completely ossified skeleton, shorter jaws, more specialized skulls, a more symmetrical tail, and thinner scales than their predecessors, and all these were evolutionary trends which were carried even further in their descendants, the teleosts, which replaced them

THE EVOLUTION OF LIFE

in Cretaceous times (Fig. 28). From then until now the teleosts have undergone an expansion of unparalleled proportions, and they are now the dominant group of both the seas and inland waters, both in number of species and individuals.

Living teleosts display an astonishing variety of form, from the salmon, herring, carp, catfish, eel, sea horse, and pike to the flounder and angler fish. They exist in countless numbers in almost every aqueous environment, from the depths of the oceans to polar and tropical streams, and even within these environments they exhibit endlessly varied habits, habitats, and diets.

It was these changes in the creatures of the seas that brought the present ocean dwellers into existence. The remoteness and inaccessibility of life in the seas tends to obscure its richness and abundance. It is only an occasional glimpse of that unseen world that reminds us of its diversity and beauty.

Chapter 16

THE DEVELOPMENT OF LIFE

July 1st, 1858, was a fine summer day in London. The temperature in Hyde Park at 9 a.m. was 70°. There was a drought and a water shortage. The Globe Theatre advertised a diorama of the Indian Mutiny. The clipper *Lonchiel* sailed for Sydney, with an estimated voyage of seventy-nine days. Parliament debated the Universities (Scotland) Bill and what W. S. Gilbert called 'that annual blister, marriage with deceased wife's sister'. Good progress was reported in laying the Atlantic Cable. Queen Victoria and the Prince Consort attended a performance of *Il Trovatore* by the Royal Italian Opera Company. And at the Oval the Gentlemen made 158 against the Players, who replied with 103 for 4, after a stand by Wisden and Lillywhite.

But one event on that day passed unrecorded in *The Times* the following morning. At a meeting of the Linnean Society, Alfred Russell Wallace and Charles Robert Darwin presented a joint paper entitled 'On the tendency of species to form varieties, and on the perpetuation of varieties and species by natural means of selection'. The paper was calmly received, and few of those who heard it could have predicted its subsequent impact upon scientific thought. Yet it was of such major importance that it marked one of the great turning points in the history of human knowledge. In 1543 Copernicus had seen an order in the endless constructions of Aristotelian cosmology, and by a unified conception of the universe had brought unity to what had before been confusion. Almost a century later, Newton reduced the order of the universe to mechanism. In the years between 1628 and 1858, although the world had seen mechanism at work in the physical sphere, the realm of living things appeared to remain distinct. It was the contribution of Darwin and Wallace to bring order and mechanism to an understanding of the organic world; a mechanism, they suggested, no less comprehensible and real than that of the inorganic world. As with the physical world, so with the organic, order had been demonstrated more than a century before. Linnaeus's *Systema Naturae* had shown the orderliness of the organic world, but the orderliness of his classification had been almost universally interpreted as the result of the special creation of subsequently immutable species.

The new theory of evolution attracted little public attention until Darwin's book *The Origin of Species* was published on 21 November 1859. The content of the book we shall discuss later (p. 284), but perhaps the most remarkable thing about the publication of the *Origin* was not the reception which it received from the scientific public, but the worldwide convulsion and outburst which it produced amongst men of all interests and persuasions. Philosophers, politicians, theologians, literary critics, historians, classical scholars, and the man in the street – all alike took it upon themselves to assess its worth. And as so varied a group studied it, so their verdicts also varied – some accepted and respected Darwin's conclusions, others viewed them with suspicion, but most rejected them out of hand, and denounced both Darwinism and all its supposed implications. 'As for the book, some treasured it, some burnt it, and some, undecided, like the Master of Trinity College, Cambridge, merely hid it!' Scientific theories, philosophies, political systems, ethical standards, revolutionary movements, social reforms, and economic *laissez-faire* – all these and more were established, modified, or justified upon Darwin's premises. Indeed Darwinism soon became all things to all men.

But the real debate focused on the theory itself. Was evolution true? Had living things evolved over unimaginable periods of time from quite distinct and different ancestral forms? These became the burning questions.

One of the unique features of the study of fossils is the opportunity that it provides to draw conclusions about the processes by which living things have developed. This does not mean, of course, that we can draw no such conclusions from the study of living things – far from it. Indeed, it is only to some

THE DEVELOPMENT OF LIFE

extent on the basis of our knowledge of the physiology and behaviour of living organisms that we can interpret some features of fossils. But it is true that it is only by examining the broad history of life that we can hope to recognize any general patterns of development which may have taken place. We have already noted some of these features and it may be useful to summarize them.

THE EXPANSION OF LIFE

Throughout geological time there has been a more or less continuous expansion of living things and this has been reflected in a number of different ways. The total number of individuals, the total number of species, and the degree of diversity which they represent have all increased throughout geological time. But within this broad expansion there has been no constant rate of increase, no common regularity of expansion, no overall trend.

Certain periods have been marked by quite spectacular expansion (the early Palaeozoic and the Mesozoic for example); others (such as the Permo-Triassic periods) by equally spectacular extinction. The same absence of regularity is seen in the increase in relative diversity. Certainly living creatures are more diverse than those of say Cambrian times, but the main patterns of animal radiation (apart from the vertebrates and insects) were established early in the fossil record. The reason for this is clear: the seas were the first environment to be colonized and the conditions of life there have remained broadly constant. It was the invasion of new environments on the land and in the air that brought new types of structure, new levels of complexity, and new ways of life, and it is these which are represented in the vertebrates and insects.

THE CONTINUITY OF LIFE

Our knowledge of living creatures makes it clear that any increase in the number of individuals can come about only by their derivation from pre-existing parents. The fossil record makes

THE EVOLUTION OF LIFE

it equally clear that the increase in diversity of living things has been brought about in the same way: that new species have developed over long periods from the cumulative modification of existing species. This is the process we saw in the horses, and all the evidence suggests that the process is absolutely continuous. There are gaps in the fossil record, to be sure, more gaps than fossils, in fact; but they appear to be gaps of non-preservation, and not of non-existence. Furthermore, the continuity extends to major groups of animals, as well as to individual species: new classes, representing totally new structures adapted to new ways of life, are developed in just the same way and with just the same continuity. This was demonstrated in the origin of the amphibia, the reptiles, the birds, and the mammals. In fact we can summarize the development of all new kinds of organisms as a process of 'descent with modification' – a formula which is the equivalent of the word 'evolution'. We hear much of evolutionary theory, but if we accept this definition of evolution as a process of descent with modification, the fossil record makes it clear that evolution is a fact.

THE INTERRELATIONSHIPS OF LIFE

No man is an island, and there is a literary, a social, and a moral sense in which we are all aware of it. There is also an evolutionary sense in which it is equally true, for no individual, no species, no way of life can ever be independent of or insulated from any other. This mutual interdependence extends to all levels and influences the whole of life. Competitors for food or space or sunlight, host and parasite, prey and predator, male and female, parent and offspring, all exist within an intricate and constantly changing equilibrium. We have seen something of its broader aspects in such things as the profound influence of the spread of mid-Tertiary grasslands on mammalian development.

THE ENVIRONMENT OF LIFE

The dependence of organisms extends to their environment no

THE DEVELOPMENT OF LIFE

less than to their neighbours. Every organism, every community is influenced by and in turn influences its environment, and a given type of life is bounded by the extent of a given environment. Fish are confined (with rare exceptions) to water, and the exploitation of the land involved a new type of life, just as the later return of reptiles and mammals to the water involved the superficial return to a fish-like body. Indeed, quite unrelated groups of animals living under similar environmental conditions tend to develop similar structures. This is the pattern of convergence or homeomorphy that we have seen in so many groups.

Now a given environment can support only a given total of living creatures. There is a definite limit to the number of sheep one can put in a pen, or plants in a garden, or even kittens in a household. Change in structure in a group of organisms can therefore be brought about in some environments only by replacement, and the rise of many new groups involves the replacement of old ones, less well adapted to the same environment. It was this pattern of change that we saw in Mesozoic plants, Paleocene mammals, bony fish, and recent South American mammals, and such replacement is a frequent feature of the development of life.

THE PERSISTENCE OF LIFE

All living things are mortal, but life is continuous. We are linked to our ancestors by a frail physical thread (the reproductive cells) that is the legacy, the source, and the bond of life down the ages. One of the most striking features of life has been the persistence of all the major phyla of living things: none has become extinct. In a few cases groups, even genera, of organisms have persisted for millions of years with no essential change, but this is the exception and not the rule. In the great majority of cases the groups to which individuals belong are no less mortal than their members, and species, genera, families, orders, and classes have all shared the common fate of extinction.

We have seen the difficulty of interpreting the cause of this extinction, yet there are a few groups from which some

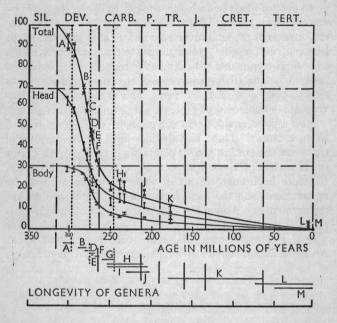

FIG. 53 Rates of evolution in the lungfish (Dipnoi). The various figures plotted are obtained from particular fossil sequences in which evolutionary change from a 'primitive' to an 'advanced' condition can be traced. Each genus is then graded and 'marked' on this scale. High marks indicate primitive conditions. The genera are represented by:

A *Dipnorhynchus*; B *Dipterus*; C *Pentlandia*; D *Scaumenacia*; E *Fleurantia*; F *Phaneropleuron*; G *Uronemus*; H *Ctenodus*; I *Sagenodus*; J *Conchopoma*; K *Ceratodus*; L *Epiceratodus*; M *Protopterus* (and *Lepidosiren*).

The graph shows rate of loss of 'primitive' ancestral characteristics and approximate time ranges of the main genera. (*After Westoll*)

282

tentative conclusions might be drawn. The South American marsupials seem clearly to have owed their extinction to the sudden influx of placental competitors from the north, and it is difficult to escape the conclusion that the great Permo-Triassic wave of extinctions was related to the unique combination of contemporary physical conditions. Certainly, in many, if not in all, cases, extinction seems to have involved organic competition, or inadaptation to changing conditions, or both.

The same is true of the closely related problem of survival. Some groups (*Lingula*, for example) seem to persist because they are eminently well adapted to a persistent environment.

Survival and extinction inevitably lead us to consider their corollary, rate of change. Survival of a group may involve no rate of change (it may remain static in form), considerable rate of change (it may survive by means of modification), or the ultimate cessation of change (it may become extinct). There are various ways in which rates of change might be expressed. Two of the most obvious are the consideration of rate of appearance of new groups per unit of time or the rate of change of some selected body structure either per unit of time or relative to some other structure. Numerous studies of this kind have been made and one of these is shown in Fig. 53.

Although at present there are so many variables and imponderables that there is little agreement in their interpretation, such studies seem to offer great promise in the future.

THE EVOLUTIONARY PROCESS

It would be inappropriate in this book to attempt a detailed consideration of the intricate and varied aspects of the evolutionary process. A companion volume in this series is devoted entirely to this one subject, and this is an indication of both its importance and its complexity. But we have seen that fossils demonstrate the fact of evolution, and it is difficult to escape the question, 'How did it come about?'.

The answer which is now generally accepted is that first suggested by Charles Darwin and Alfred Russell Wallace in

THE EVOLUTION OF LIFE

1858, and later amplified and justified in Darwin's great book *The Origin of Species* (1859). The essentials of Darwin's argument are simple (and from our retrospective viewpoint, obvious). Darwin claimed that living species have arisen from pre-existing species by a process of descent with modification. Now this suggestion was not new. Other naturalists (Lamarck, Buffon, and Erasmus Darwin among them) had suggested descent with modification long before this, but the scientific world remained sceptical, and generally hostile. It was not until the eighteenth century that the true nature of fossils had been generally recognized, and its interpretation was governed by the widespread acceptance of Archbishop Ussher's chronology, which by detailed calculation from the Genesis narrative reckoned creation to have taken place in 4004 B.C. at 9.0 a.m. on 26 October. Because of this general climate of opinion the 'orthodox' view of creation had come to be based upon a theory of catastrophism, which maintained that the earth had experienced a number of successive cataclysmic revolutions (of which the Noachian deluge was the most recent). Each of these catastrophes was thought to have completely destroyed all living things, so that after an interval of time a new creation took place, whose beings were in turn entombed in the strata of the next cataclysm.

It is easy to smile at what we now regard as such a naïve concept, but it held undisputed sway in the scientific world for over a century, and found for its champions many of the greatest pioneers in the development of the natural sciences.

The importance of Darwin's work lies largely, therefore, not in the fact that he was the first to suggest the possibility of evolution, but in the fact that he convinced the great majority of scientists that evolution had taken place, and this he did by the presentation of a wealth of detailed data which supported his theory of the mechanism by which evolutionary changes had been effected.

The evolutionary hypothesis of Darwin and Wallace rested on three essential foundations. Two of them were observations, and the third an inference. They noted, first of all, that all

THE DEVELOPMENT OF LIFE

organisms tended to over-produce. This proliferation of offspring is something that seems characteristic of the whole world of living things. A far greater number of young are produced than ever survive to maturity, because on the whole the number of individuals within particular species remains more or less constant. Many calculations of this super-abundance of nature have been made. The oyster, for example, produces something like 600 million eggs per season, and we are told that if the great-great-grandchildren of one such group all survived without mortality, and reproduced, their shells would number 66×10^{33} and would make a mountain eight times the size of the earth.

Secondly, Darwin and Wallace observed that individuals in any species are not identical, but show variation in certain characteristics, and that, although some variation is spontaneous, in many cases variations (however produced) may be inherited. The more obvious of these variations include such things as size, speed of movement, health, fertility, instincts, physiological efficiency, and so on. The mechanism of this inherited variation was unknown to Darwin and Wallace, but since their day genetic studies of inheritance and mutation have provided the basis of an understanding of the process. From these two observations Darwin and Wallace concluded that some of these variant forms must inevitably stand a better chance of survival than others, and they would tend to produce relatively more offspring than those less fitted for the environment in which they lived. Therefore, on the whole, reproduction would be non-random, the better adapted producing relatively more offspring, whose characters would include those advantageous ones which the parents possessed. This was natural selection, the mechanism by which Darwin and Wallace sought to explain the vast diversity of living things.

Although Darwin stressed the fitness of organisms to their environment, and the drastic changes produced by domestic breeding, natural selection remained a hypothesis. It is now a fact and has been convincingly demonstrated in a number of living forms. A particularly good example is the work of H. B. D. Kettlewell on industrial melanism in moths. Until about the middle of the last century, the British peppered moth, *Biston betularia*, existed commonly in its typical light form, having a grey background with pepperpot markings on the wings. A very rare, dark melanic variety was also known as *carbonaria*. This variety is controlled by a single dominant mendelian gene, and is slightly more vigorous than the normal grey type. Because of its conspicuous colour against the lichen-covered trees, this dark form, wherever it appeared, was constantly eliminated, and it made up only about one per cent of the populations in which it occurred. It persisted only because of the repetition of the same mutation. The spread of industrial pollution within the last century has brought about a change in the environment, and under these conditions the *carbonaria* variety has increased in numbers and has proved to be physiologically 'hardier' than the peppered form. It now makes up about 99 per cent of the total population in industrial areas, and it is now more black than it was a century ago. Kettlewell has shown that the rate of predation of this moth by various birds corresponds well with the relative frequency of the two forms in different environments. The light form still persists and is still dominant in areas unaffected by industrial pollution. This is an elegant example of natural selection in action; in a very short period of time, pre-adaptation in the form of the melanic variety has been encouraged by the spread of a polluted atmosphere. The mutant appeared considerably later in some parts of Germany. Industrial melanism has now been described in about sixty other species of moth in western Europe and the United States, but it affects only those species resting on tree-trunks or other exposed positions.

It would be wrong, however, to suggest that natural selection acts simply as a bludgeon-like extinguisher. The subtlety of its action has been well shown by the work of Cain and Sheppard, for example, on the survival rate of snails with slightly different colour banding patterns on the shell. In this case selection depends very much on the season, one particular colour being better suited to one season of the year than another, but being relatively less well suited at other times. This raises an important point with regard to the whole question of natural selection. It is, in fact, always a compromise; a compromise between different seasons, between different aspects of the habits of the organism, between different parts of the organism, between different members of the population, between the individual and the community. There are many other examples of natural selection that might be quoted, a problematical one, so far as we as human beings are concerned, being the growing resistance of a number of bacteria to antibiotics. It has recently been shown, for example, that the bacterium *Escherischia coli* has become resistant to streptomycin, not because of any gradual increase in resistivity but because of a mutation which is viable in a streptomycin medium. Mutations are already there, and once a favourable environment occurs, they are selected.

But this mechanism of natural selection presents problems in the fossil record. Can we also observe natural selection in the fossil record? It is, of course, quite impossible to demonstrate small changes in colour in fossils, but we can find evidence of other comparable changes. The evidence at present is limited, but it exists. Let us consider a single example. Kurten has made a study of the European cave bear, *Ursus spelaeus*, from the Pleistocene of Odessa, which he compared with skulls of the living bear, *Ursus arctos*, from Finland. In both species there is a positive correlation of height on length for the second molar, yet the average relation between the two dimensions is identical. The fact that the two species have a similar index of tooth height to length suggests that this shape is optimal for the tooth crown in question, although it must have been produced by a genetic change within the earlier history of one or both species. The two allometries may be traced back to the Lower Pleistocene common ancestor, *Ursus etruscos*, but the similarity could not have arisen without the influence of a genetic change, because continuation of existing trends would have produced dissimilar indices.

Kurten has been able to go further than this, however, and to show the way in which certain differences appear to be selected by environmental conditions. He studied a large fossil sample from caves near Odessa, and the results indicate both the type and amount of information which detailed fossil analysis may now provide. There is a marked separation of younger individual fossil bears into a series of discrete stages of development. This is seen in samples from many localities, and the stages are virtually identical. This is best interpreted as the result of the caves from which collections were made having been inhabited only during hibernation, so that the growth stages are separated by a period of a year. The youngest specimens, equal in size to new-born cubs of the related living species *Ursus arctos*, are rare, presumably because of the fragility of the bones. The predominant ontogenetic stage is one in which all permanent teeth are already formed, but in which only a few of them are in place. Comparison with living bears shows this to represent an age of four to five months. The cubs were evidently born during the period of hibernation, and the preponderance of specimens of this stage, with only very slight variation in both directions, indicates a mortality peak of short duration towards the end of hibernation. The succeeding growth stage represents an age of about 1·4 years, indicating the absence of the bears from the cave until the next hibernation. Growth stages for successive years show that at about four years the bears are fully grown. Sexual dimorphism, although traceable back to the first years of life in the canines, becomes pronounced for other structures in the third year, which is thus apparently the period of puberty.

The rate of mortality for each of the age groups may be calculated by dividing the total number of teeth in a given age group by the sum of these and all older homologous teeth. After careful allowance for bias and difficulties of interpretation in older specimens, Kurten constructed a life table (Fig. 54). This is of sigmoid form and shows a striking resemblance to similar tables for other groups, including man. It was further tested by comparison with other cave collections and with living bears. It includes decreasing mortality in young individuals, stable rates for those in their prime (five to ten years), and increasing rates for the old. Life expectation at birth was about three and a half years and the maximum age eighteen years.

Successful hibernation results mainly from adequate feeding during the preceding season, so that bear teeth are especially suited for studies of natural selection. Kurten studied the index 100 (length of the largest cusp (the paracone) of the second molar/total length of the second molar). This index shows a

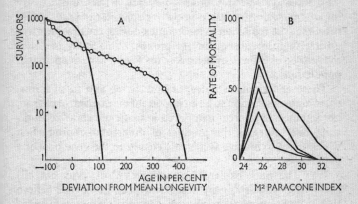

FIG. 54 Natural selection in Pleistocene cave bears.
A Survivorship curves from birth for *Ursus spelaeus* from Odessa and *Homo sapiens* (white males, continental U.S.A.); B Centripetal selection with a linear component simultaneously acting on the relative paracone length of the molar M^2 of *Ursus spelaeus* from Odessa. For explanation of figure see text. (*Figure redrawn after Kurten (1958), with the permission of the author.*)

marked change in mean and standard deviation with age, the standard deviation showing a slow but steady reduction in variation with age. The change in the mean index is highly significant, and its continuous reduction with increasing age is a clear indication of natural selection in favour of a smaller than average paracone. This may be expressed graphically (Fig. 54) and indicates strong natural selection in favour of a smaller than average paracone, with an optimal index of 26 (a skewed point within the range). This selection started at a very early stage, as soon as the teeth became functional; within two or three years the inadapted had been eliminated, and the subsequent standard deviation of the sample reduced. Other tooth dimensions were tested and gave comparable results, which were explicable in terms of functional efficiency. Samples from other localities gave broadly similar, but not identical, results, suggesting differential selection in varying microenvironments. This intensive selection of an apparently trivial, but actually important, character was especially strong in the years before breeding (the reduction in mean index during the first year was greater than in subsequent age groups). This study indicates the powerful effects of selection on fossil, just as on living, populations.

These and similar studies, of both living and fossil forms, show natural selection to be an important mechanism of change, not so much as a fierce 'nature red in tooth and claw' relationship, but rather as the product of differential reproduction which tends to involve systematic change in the gene pool of a population.

Was Darwin's explanation correct? Most (but not all) zoologists, botanists, geneticists, and palaeontologists believe that it was, although it is now accepted with some minor modifications. The fact of evolution as such has been chiefly, but by no means exclusively, demonstrated by the study of fossils, but it is not surprising that it is the study of living rather than fossil forms which has provided most information about the process of evolution. Certainly, however, the fossil record confirms both the abundance of life in the past, and the universality of infra-specific variation. Furthermore, in the very few cases where it is possible to draw any general conclusions (in problems of extinction, survival, competition, replacement, and so on) the fossil evidence seems to support Darwin's theory. What the fossil record has not yet provided, and never will, is an unbroken and detailed record of the history of life, a series of frame-by-frame stills of fossil organisms, so complete that they may be played back to provide a continuous film of evolution. This should not surprise us.

Darwin was conscious that the fossil record represented the

THE DEVELOPMENT OF LIFE

only opportunity to study the course of evolution and, as he viewed it, he acknowledged that far from providing a continuous cine-film of the development of life it consisted of a series of gaps. Now our present knowledge of the fossil record is very different from that of Darwin. The fossil record is known in far greater completeness than it was in his day. How then are we to read the fossil record? Our biggest concern is still with the problem which faced Darwin. In brief, does the fossil record show evidence of continuity, or of discontinuity, in the rise of new species and higher groups? Although most palaeontologists now regard it as indicating an imperfect sample of a once essentially continuous evolutionary sequence, a small number of distinguished workers regard the admitted 'higher' gaps in the fossil record as inherent in the evolutionary process itself, rather than as geological in origin. It therefore seems worth while to examine this key question.

Let us first summarize the ways in which the fossil record shows evidence of continuity. We need review this only briefly, for palaeontological literature provides many such traditional evolutionary 'proofs'. The first, and most obvious, way is that it shows a general progression from what we now regard as 'simple and primitive' forms to more 'complex'. There is also throughout the history of life an apparent expansion in numbers of species, in diversity, in the total volume of the biosphere, and in the independence of some organisms from their environment. This gradational, successional, 'progressive' character of the fossil record is so well known that it is often overlooked, but it is of major importance in evolutionary thinking. The expanding, diversifying, environment-expanding, replacing development of living things is just what one might predict *a priori* on the basis of evolutionary theory.

Secondly, at all taxonomic levels, there are now, in a limited number of cases, examples of continuity. Let us first of all take high taxonomic levels. We have already seen, especially in the vertebrates, remarkable transitional forms between various classes. Between the crossopterygian fish and the amphibia, we have the ichthyostegids, part fish, part amphibia, known from the

Upper Devonian or Lower Mississippian of Greenland (p.178). Between birds and reptiles, we have the renowned *Archaeopteryx* (p.218). Between amphibia and reptiles, we have the seymouriamorphs (p.126). Between reptiles and mammals, we have the therapsids (p.196). We have, in the vertebrates at least, good evidence of complete continuity between classes. It may not be a matter of pure coincidence that in the vertebrates the classes were established much later in geological time than in other groups. The majority of invertebrate classes originated in either Pre-Cambrian or early Palaeozoic times. Could the lateness of the vertebrates' establishment have favoured their preservation?

Secondly, at a lower taxonomic level, between genera, for example, we also have a substantial number of transitional sequences. One of the best of all is the sequence of horses linking the whippet-sized, primitive, Eocene form, *Hyracotherium*, with the living horse. This was one of the first fossil sequences ever described. It was first described by Kovalevsky in 1874, and it was later amplified by Marsh, and interpreted by Huxley. The beautiful gradational sequence which these fossils show has already been described (p. 239, Fig. 41), so that we need only summarize its major features. These involved the increase in body size, the increase in size and change in the shape of the skull, changes in the teeth, involving the pre-molarization of the molars, and the deepening of the teeth from low crowned to high crowned, together with the infilling of the depressions in the upper surfaces with cement. With these were associated changes in the limbs, with the gradual reduction in the number of toes, and in the whole change in construction of the limbs associated with the change in posture from pad-footed to spring-footed. Now this series is incontrovertible. It provides clear evidence of the transition of one genus to another over a period of something like 70 million years, and it is a series that could be paralleled in many other groups; in the titanotheres, in the ceratopsian dinosaurs, in the proboscideans at the one end of the scale, and in the protozoans at the other, as well as in many other groups.

Thirdly, there is evidence in a number of cases of continuity between species. In almost any species group studied in detail, there is evidence of gradual transition. Micropalaeontologists tend to work with samples which are particularly well suited to demonstrate this. We might take one such example. Kennett has described evolution in the Upper Miocene and Lower Pliocene *Textularia miozea-kapitea* lineage (Fig. 55). His specimens – from the Cape Foulwind area, New Zealand – were collected over a stratigraphic interval of some 450 feet and show a progressive decrease in width-depth ratio. Taxonomically they are interpreted as representing both intra- and infra-specific evolution. Although the sizes of some of Kennett's samples seem inadequate, his general conclusions appear to be valid. Many comparable microfaunal examples could be quoted.

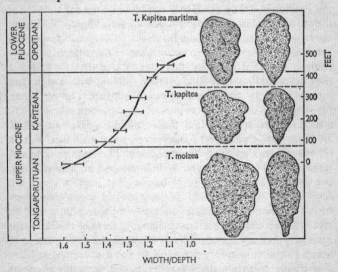

FIG. 55 Evolution in Tertiary foraminifera.
Curve showing average width-depth ratios for successive population samples of the *Textularia miozea-kapitea* lineage from the Upper Tertiary of Cape Foulwind, New Zealand. The standard deviation for each sample is shown, and representative individuals illustrated (*redrawn and modified from figures by Kennett (1963) with the author's permission*).

So much then for the fact of continuity between various taxonomic levels. It exists in many groups, and, although a few classic invertebrate evolutionary series are at present in dispute, the evidence for continuity at all taxonomic levels is clearly demonstrated. We shall have to ask later whether we may regard it as typical of all new taxa.

We turn now to the second aspect of continuity, which is its nature, and here there are two particular features that we need to consider. The first is one that we can illustrate very well by looking at the example of the transitional forms which link one class of vertebrates to another, of which the ichthyostegids are an example. We have seen (p. 178) that they provide a classic example of what has come to be known as mosaic evolution. The ichthyostegids are not themselves the actual ancestors of the amphibia – they are rather too late in time for that – but they must be almost identical to the ancestral forms. Although they are intermediate in many features between crossopterygian fish and amphibia, they are not uniformly transitional in every respect. If we analyse them character by character we find that they are a jumble, part fish, part amphibian, with a complete intermixture. In some senses they are still fish. The vertebrae, for example, are highly fish-like. The median fin and tail structures are quite unlike those found in later amphibia. In these two characters, ichthyostegids are still fish. But in other respects they are fully amphibian. The strong pelvic and pectoral girdles, the limbs strongly developed for land walking, the double-headed (bicipital) ribs are all characters which are found only in amphibians. So they are, in some characters, fully fish; in other characters, already fully amphibia, and in still other characters, such as the general form of the skull, for example, they are about mid-way between the basic pattern of bones in the crossopterygian skull and that of the amphibian. The basic structural pattern is still retained, but some elements are reduced and others suppressed; the eyes are closer together than they are in fish, and so on.

Now this jumble of characters, some more fish, others more amphibian, is found also in every other transitional form. We find it in *Archaeopteryx*, we find it in the seymouriamorphs, we find it in the therapsids, and we find it also at other taxonomic levels. We find it in such living examples as the duck-billed platypus. We find it in *Homo erectus*, the ancestral form from which later hominids probably evolved. It is a feature which is strongly suggestive of the efficacy of natural selection, because one of the arguments which has frequently been made against natural selection is that it is quite impossible to conceive of it producing the closely co-ordinated set of changes that would be required to convert, say, a reptile into a bird. But, as we have already seen (p.56) in the case of *Archaeopteryx*, we find that there was no such co-ordination of all character changes, but that the various intricate changes involved took place not at a uniform or similar rate but rather randomly, one at a time, so that the whole bird was not produced until long after some characters of the 'reptile' were already fully birdlike. The argument has been made that it is inconceivable that natural selection could have gradually transformed arms into wings, provided a sternal keel and girdle, aerated, light, strong bones, warm blood, and feathers, and controlled all the other essential structural and physiological modifications that we associate with birds. But when we look at *Archaeopteryx*, we find that it shows no such co-ordination. We have seen, for example, that the breast bone, the sternum, of the creature lacked a keel. It was poorly ossified and spongy in texture, and the clear implication is that *Archaeopteryx* lacked the strong breast muscles associated with flight, apparently gliding rather than flapping its wings very actively. This conclusion is confirmed by the structure of the brain, for De Beer, using ultraviolet photography, showed that the small size of the cerebellum was quite inappropriate to control the balanced co-ordinated movements required by a flying bird. Thus the evolution from reptiles to birds took place in a series of steps which did not demand the intricate synchronization of characters which the critics of evolution supposed. There are other ways in which *Archaeopteryx* is a jumble of characteristics, but it is sufficient here to repeat that the characters of birds appeared, not all together, synchronized and co-ordinated, but one by one, slowly, irregularly, piecemeal, and there existed no totality of bird characters as we define them until the last had been added a very long time after the first. In this, and in all such transitional forms, we have strong evidence of natural selection at work.

The example of horses, which we discussed earlier, provides a second striking illustration of the nature of continuity. One of the things that Darwin himself pointed out was that, in general, if we look at forms which are very distinct, we have no reason to suppose that the links that once existed between them were directly intermediate between each and an unknown common ancestor. He took as an example the horse and the tapir, and he pointed out that we should probably be unable to recognize the parent form of these two, even if it were to be found. Now it so happens that when Darwin wrote this in 1859, an animal very near the common parent of horses and tapirs was actually known. But in structure it was far from being intermediate between the two; it was very different from both. It was not recognized as the common ancestor until the transitional sequences of intermediate forms leading to horses and tapirs were discovered long afterwards. The ancestral form was *Hyracotherium*, described and named in 1840. Its very name, *Hyracotherium*, was given by Owen because of the similarity of its teeth to those of living conies or hyraxes. Although, as Simpson has pointed out, the contemporary form, *Homogalax*, may be slightly closer to tapirs, these two are so similar in structure that palaeontologists still make mistakes in distinguishing them.

Although there are examples of continuity at all taxonomic levels, we are led to ask why continuity is not greater than this, for these examples are not typical of the fossil record. There are more gaps than sequences. We have, on the whole, very few sequences between the classes; we have numbers of transitional sequences between genera, and relatively few between species. Why do these gaps exist? Are the gaps biological or essentially physical in origin? Is normal change gradational but the record imperfect, or is the small number of described transitions atypical of the process of change as a whole?

Why is it that the record contains far more gaps than it does transitional sequences? There is no single, simple answer to this question but there are a number of aspects that suggest a variety of reasons. The first and most important of these we have already briefly discussed. The fossil record is both incomplete and biased. One example of this incompleteness must suffice. It concerns the evolution of horse genera, which we have already discussed. Morley Davies analysed the number of fossiliferous horizons in the western United States which had yielded fossil horses. He pointed out that in the sixty to seventy million-year period of history which the family represented, only eighteen fossiliferous horizons were known when he wrote. If we assume by comparison with living horses that the length of generation in fossil horses averaged four years, this period of time would represent something like fifteen million generations. Thirty species are involved in the transitions represented, each having an average time span of two million years, and comprising some 500,000 generations. For this vast span of time there were eighteen fossiliferous horizons. Recorded human history represents only 5,000 years and 200 generations. Even though many more fossiliferous horizons are now known, this example remains a striking indication of the inadequacy of the fossil record.

Still other gaps in the fossil record are clearly the result of inadequate collecting and study. Most of our palaeontological collecting is not directed towards the acquisition of evolutionary series and most palaeontologists have not the time or the interest to trace the many individuals involved in such sequences. It may be significant that a growing proportion of the new evolutionary sequences are being described in microfaunas and a large proportion of the existing sequences are in vertebrate faunas. In the first case the samples tend to produce very large numbers of individuals, and in the second the zoological training and motives of most vertebrate palaeontologists make them more interested in the evolutionary implication of their fossils than their invertebrate colleagues. (The use of the latter term in this context is not to be taken literally, and implies no disrespect.)

Furthermore, many faunal sequences are interrupted by migration. The fossil record contains striking illustrations of the importance of this as an explanation of discontinuity. Simpson has shown the effect of this in phylogenetic reconstruction of horses, based respectively on faunas collected from North America and Eurasia. In the latter case the few known fossil representatives could suggest saltationary development, but the completely transitional North American sequences show the 'discontinuous' European series to be occasional migrants from the main evolutionary stream. Migration is especially common in times of radical environmental transition, where the new descendant group has become adapted to an entirely new environmental zone. Often the ancestors, and sometimes the descendants, are geographically restricted.

This is especially well seen during the later evolutionary phases of certain groups. For example, the coelacanths were a widespread and diversified group in Palaeozoic times, but they were generally supposed to have become extinct in the Cretaceous. Recent representatives are now known from a number of areas off the coast of East Africa, however, and their disappearance from the fossil record for a period of something like seventy million years must be explained by their restricted distribution. An even more striking example of the absence of fossil representatives is provided by the monoplacophoran molluscs, thought to have been extinct since Devonian times. A representative of a living species of this group, *Neopilina galathea*, was dredged up from a deep off South America in 1951. Similarly, the trigonids have a widespread history throughout most of Mesozoic time in many parts of the world, but their post-Mesozoic history is entirely confined to the southwest Pacific. They became extinct elsewhere at the end of the Cretaceous, and living forms are entirely restricted to a handful of closely related species of the genus *Neotrigonia*, distributed around most of the Australian coastline. They are, in fact, though geographically restricted, a very ecologically unrestricted

THE DEVELOPMENT OF LIFE

and tolerant group, occurring in seas ranging from tropical to cool-temperate, and in depths ranging down to 200 fathoms, a tolerance shown by few other pelecypod genera.

The same feature is seen in the late history of blastoids, which became extinct in early Pennsylvanian times in most parts of the world, but lingered on in Timor until the Permian. The implication of these and other such examples is that fossilization favours abundant and widespread groups. Both the late representatives of some particular groups and early ancestral forms of most groups are probably neither common nor widespread and this must be a major zoological, as opposed to purely geological, reason for discontinuities in many critical sequences. Such examples also provide clear demonstration that, at least in these cases, the assumption that absence of intermediate fossils is indicative of the non-existence of intermediate forms is invalid.

It is probably also true that the most rapid evolution often takes place in relatively small and isolated populations and this again reduces the chance of subsequent fossilization. All this may well be considered to represent very special pleading. Yet it is a fact that in the fossil record there are repeated demonstrations that great evolutionary episodes of diversification have proceeded from very limited ancestral stock. Carboniferous ammonoidea, for example, included only two Devonian survivors, Triassic ammonites arose from two superfamilies, and the hordes of Jurassic and Cretaceous ammonites from only one genus, though in these cases the ancestral stocks were relatively widespread.

There are two other aspects of the fossil record that provide strong circumstantial confirmation that what we observe in it is a sample of a once-continuous evolutionary chain. Simpson has shown that it is possible to construct an experimental model, using either an inferred or a hypothetical evolutionary phylogeny, in which the species are numbered, and then segregated into genera and families. A random sampling is made of, say, 10 per cent of these numbers, and the results tabulated. Although such a sampling makes no allowance for the intrinsic and imposed bias of the fossil record, the sampling pattern that emerges contains very few sequences between families, relatively few between species, and rather more between genera. The size of the 'gaps' (measured by the number of 'missing' species) increases markedly as the taxonomic hierarchy is ascended. That is, the sample pattern, obtained by testing the hypothesis that the fossil record corresponds to a once-continuous phylogeny, correspond closely with those obtained in practical palaeontology. There are presumably rather more transitions between genera than between species because we need only one species of polytypic genera to form transitional generic sequences, whereas we need particular species to provide specific phylogeny. The whole device of taxonomic categorization makes it generally the case that the percentage knowledge of transitional fossil species will always be more influenced by inadequate collecting than will higher taxa. A small sample of species represents a relatively higher proportion of the once-living total of each successively higher category.

A second aspect of the discontinuities within the fossil record is of major importance. One of the best-established features of the development of any scientific hypothesis is that, if it can be used successfully to predict future events, we regard it as valid. Its elevation to a theory, and ultimately to a law, depends on its repeated continued success in prediction. This is one standard method of verification. A very good example of the way in which evolutionary palaeontological hypothesis has been used to make such predictions is the case of the ancestry of the reptiles. When D.M.S. Watson delivered the Silliman lectures at Yale in 1937, he predicted the form that he would regard as ancestral to the reptiles. It was not then known, and his description was simply a prediction made by extrapolating backwards the known evolutionary tendencies of early reptiles. Now, because of World War II, the lectures were not published until the 1950s, but by then such a creature had been found by Romer and Whittier. In every part of its essential morphology it agreed with Watson's prediction. It was too recent geologically

THE DEVELOPMENT OF LIFE

to be the direct ancestor of reptiles, but it must have been very similar in structure to the forms that were.

Another such example is provided by the monoplacophoran molluscs. Before the discovery of a living member of this supposedly extinct group, Knight had reconstructed a molluscan phylogeny, in which fossil monoplacophorans were shown as the ancestral stock for chitons and also for other gastropods. The discovery of the living fossil *Neopilina*, off the coast of Costa Rica, published in 1957, proved the former existence of the hypothetical ancestral mollusc constructed five years before, and led to the recognition of a new molluscan class.

It would be wrong to suggest that support for the Darwinian theory of evolution had come only from the fossil record, however. Intensive studies of living organisms and processes have produced new data of great significance.

One of the problems that baffled Darwin, for example, involved the question of the origin and inheritance of variations in animals and plants. Darwin believed, like almost all his contemporaries, in a system of blending inheritance, and, to some extent, in the inheritance of acquired 'characteristics'. We now know that Darwin's views were incorrect, and in fact during the lifetime of Darwin, Gregor Mendel, an Austrian monk, who was studying the growth of peas, provided an answer to the question of the mechanism and laws of inheritance.

We now know that the basis of the inheritance of particular characteristics and particular offspring is to be found in the gene, a name which we use to denote a self-copying unit of protoplasm that is common to virtually all groups of animals and plants. These genes are arranged in a definite order within visible cells called chromosomes. Recent work has thrown new light on the way in which the pattern of replication is produced, and especially the way in which the detailed coding of information is carried within the DNA molecular structure.

We know now that the major source of variation in living things is to be found in the re-shuffling of genes which takes place in the reproductive cells, and which are then recombined when fertilization takes place. We can therefore see that on the

THE EVOLUTION OF LIFE

one hand definite linear pattern of genes gives us the continuity of character which is so characteristic of any single species of animals, but the re-shuffling or sexual recombination of genes at reproduction gives us, within that broad similarity, an obvious variability.

There is, however, one other major source of variability in living things. Genes are highly complicated structures containing many thousands of individual patterns. Because of this complexity the process of self-copying is not always exactly perfect. Sometimes, very occasionally, the copy which is produced shows some difference from the original. This imperfect copying of genes is spoken of as mutation. Whereas the sexual recombination and shuffling provides new variations and combinations from existing genes, the importance of mutation is that it provides completely new characteristics and variations. This is an enormously complex process, and in one sense we do violence to it by attempting to describe it in such broad terms as those that I have used. Even the tiny fruitfly, *Drosophila*, is estimated to possess about 5,000 genes, each one different, each one containing many thousands of atoms. This complexity of life then is enormous, and as we consider the course of evolution as represented by the fossil record we have taken a grossly superficial view. That is not to say that it is an improper or an unrewarding or an unproductive view. It is rather to say that it is only one way of approaching the vast problem of the evolution of living things. At the other end of the scale, at the level of electron microscopy, of biochemistry, of genetics and biophysics a second major attack is now yielding the most exciting information about the mechanism of inheritance.

In other ways, too, the study of living organisms confirms and illustrates Darwinian evolutionary theory. In spite of the bewildering abundance and diversity of life, biochemical and physiological studies show that the one and a half million species of organisms all solve the basic problems of survival in much the same way. Their basic composition, cellular structure, and metabolic processes are all essentially similar. All employ nucleic acids as the basic unit of heredity, for example, and the

THE DEVELOPMENT OF LIFE

same particular phosphate (ATP) in energy transfer. Furthermore, their individual developmental patterns, including those of their embryos, degrees of similarity, structural patterns, geographic distribution and interdependencies are most coherently and economically explained in such evolutionary terms. Degrees of biochemical and serological similarity, based, for example, on protein composition and immunity reactions, show an astonishingly close correspondence to the classification of organisms, based on quite different data and assumptions. This implies that the classification of organisms is not merely an arbitrary mental construct, but reflects in varying degrees ancestral–descendant relationships.

So studies of living species support those of fossil forms in suggesting the adequacy of the Darwinian evolutionary recipe. Neither provides strict proof, in the sense that some opponents of evolution have sought it. Yet both living and fossil organisms suggest that the history of life, the course of evolution, is consistent with a simple mechanism, although this does not imply that the course of evolution is itself simple. The striking thing is that in such an almost inconceivably long and complex history, involving hundreds of millions of species, we can detect any pattern at all. We should be concerned at the naïvely simple ways in which we sometimes tend to approach and reconstruct evolutionary patterns. Certainly we impose our laws upon, rather than extract them from, the data of the natural world, and there are real dangers in our oversimplification. Such a complex result does not demand a complex process, but perhaps our greatest palaeontological hazard is to allow ourselves to become so absorbed in our unified explanation of the evolutionary mechanism that we fail to give adequate consideration to new or conflicting data. The common property of scientific models is that they ultimately become discarded in favour of others which account for the same or additional phenomena with greater economy or precision. Neo-Darwinianism may ultimately suffer a similar fate, but at present it provides an economical and elegant explanation for the long course of evolution.

THE EVOLUTION OF LIFE

Yet it is not easy to develop the mental elasticity to comprehend the endlessly complex history of life in such terms. There is, however, one other way in which the fossil record provides evidence of the correctness of the theory of natural selection, for it is sometimes argued that this process is inadequate to account for the endless diversity of life and the enormous differences in levels of organization which it represents. Can it really be so simple as this? Can an odd mutation, a minute variation, a slight advantage, be adequate alone to effect the changes the earth has seen? Can one seriously believe that so feeble a mechanism can produce such mighty results? The lowly life of the seas that bore the first living creatures – the hosts of shell-bound invertebrates that thronged the shores of ancient lands, the darting shoals of armoured fish that marked the advent of a new level of complexity in body structure, the clumsy squat amphibia that recorded the first tenuous conquest of the land by vertebrates – can it really be that variation and selection, these mechanisms alone, can produce such wonders as these?

We are tempted to deny it, but one other factor has to be assessed and that is the factor of time. With the help of time, adequate time, an eternity of time, what then?

Early estimates of the age of the earth proved a serious embarrassment to evolutionists. Lord Kelvin in the 1890s suggested, for example, that the earth could not be more than 20 to 40 million years old, and this seemed a quite inadequate period of time to allow the great changes in form of living things to have arisen by a process of slow continuous change. We have already seen (p. 74), however, that present estimates of the age of the earth suggest a figure of 4,500–5,000 million years, and some Pre-Cambrian fossils appear to be at least 3,200 million years old. Such a span of time is of enormous importance to the theory of evolution for it allows ample time for the postulated slow process of change to have taken place.

Time studies also permit a tentative assessment of evolutionary rates, as we have seen, and provide an estimate of the time involved in the transition of one species to another. The average

THE DEVELOPMENT OF LIFE

time involved in speciation during the adaptive radiation of the mammals was probably of the order of 500,000 years, while the average rate of change in most morphological characters, even in 'rapidly evolving' groups, is very slow. Simpson has shown, for example, that the average change in diameter of early equid molars is less than 0·2 mm. per million years, while differences of 3·0 mm. or more were present within single populations. These rates of change are thus so slow in most groups that it is generally quite impossible to observe such 'natural' (as opposed to artificial) change in contemporaneous faunas. This readily answers those who complain that no one has ever seen a new species develop. Yet, slow as these rates of change are, the length of time over which they have occurred is so great that even such imperceptible changes marked the origin of new forms of life.

EPILOGUE

WE have traced the steady development and growth of life from its hazy origin, and early manifestations, down the long corridors of time that lead to the present. It is a wonderful story, a history of millions upon millions of millions of individuals, of millions of different species, through thousands of millions of years. There is nothing more breathtaking in the whole of human experience than the contemplation of this ceaseless cavalcade of living things, hovering between birth and death on the surface of our tiny planet. In an endless procession animals and plants have spread and multiplied and vanished, each for a fleeting moment a part of the continuing process we call life. But this outer view is only one facet of the mystery and wonder and beauty of it all, for within and through and around the bodies of each of these countless individuals there has pulsed the breath of life – at a score of levels of complexity, a bewildering maze of intricate chemical and physical and biological changes have interacted together to produce and maintain this frail thread of life which binds us all. Yet 'he who stops at the fact misses the glory', and if in the search for the pattern and process of its development we miss the wonder of life and fail to grasp its deep significance for our thinking, then we have missed the glory.

But the man who has caught a glimpse of life in this perspective can only see it as a thing of reverence and wonder. How life evolved we now begin to understand; that it evolved remains a source of wonder; why it evolved is not a question with which science as such is concerned, but it is a question that links the other two, and gives to the scientific quest a purpose and a harmony within the broader and deeper unity of human contemplation. And it is ultimately upon our answer to this question that we build the fabric of our own portion of the history of life.

SUGGESTIONS FOR FURTHER READING

GENERAL BIOLOGY

Buchsbaum, R. *Animals without Backbones*. Harmondsworth (Penguin Books), 1951.
Romer, A. S. *Man and the Vertebrates*. Harmondsworth (Penguin Books), 1954.
Simpson, G. G., and Beck, W. S. *Life*. New York (Harcourt, Brace and World), 1965.
Yonge, C. M. *The Sea Shore*. London (Collins), 1966.

EVOLUTION

De Beer, G. R. *Atlas of Evolution*. London (Nelson), 1964.
Moody, R. C. *Evolution*. New York (Time-Life, Inc.), 1964.
Rhodes, F. H. T. *Evolution*. New York (Golden Press), 1974.
Simpson, G. G. *The Major Features of Evolution*. New York (Columbia University Press), 1953.
Smith, J. M. *The Theory of Evolution*. Harmondsworth (Penguin Books), 2nd edn, 1966.
Evolution. London (British Museum of Natural History), 1958.

PHYSICAL GEOLOGY

Gilluly, J., Waters, A. C., and Woodford, A. O. *Principles of Geology*. San Francisco (W. H. Freeman & Company), 1968.
Holmes, A. *Principles of Physical Geology*. London (Nelson & Sons), 1965.
Longwell, C. R., Flint, R. F., and Sanders, J. E. *Physical Geology*. New York (Wiley & Sons), 1969.
Rhodes, F. H. T. *Geology*. New York (Golden Press), 1972.

HISTORICAL GEOLOGY

Cloud, P. C. *Adventures in Earth History*. San Francisco (W. H. Freeman & Company), 1970.
Dunbar, C. O., and Waage, Carl. *Historical Geology*. New York (Wiley & Sons), 1969.
Kummel, B. *History of the Earth*. San Francisco (W. H. Freeman & Company), 2nd edn, 1970.
Moore, R. C. *Introduction to Historical Geology*. New York (McGraw-Hill Book Co.), 1958.

SUGGESTIONS FOR FURTHER READING

Raynor, D. H. *The Stratigraphy of the British Isles.* Cambridge (Cambridge University Press), 1967.

Woodford, A. O. *Historical Geology.* San Francisco (W. H. Freeman & Company), 1965.

PALAEONTOLOGY: GENERAL

Beerbower, J. R. *Search for the Past.* New Jersey (Prentice Hall), 1969.

Black, R. M. *The Elements of Palaeontology.* Cambridge (Cambridge University Press), 1970.

British Caenozoic Fossils; British Mesozoic Fossils; British Palaeozoic Fossils. London (British Museum of Natural History), 1959.

Brower, A. *General Palaeontology.* Edinburgh (Oliver & Boyd), 1967.

McAlester, A. L. *The History of Life.* New Jersey (Prentice Hall), 1968.

Oakley, K. P., and Muir-Wood, H. M. *The Succession of Life through Geological Time.* London (British Museum of Natural History), 1956.

Rhodes, F. H. T., Zim, H. S., and Shaffer, P. R. *Fossils, a Guide to Prehistoric Life.* New York (Golden Press), 1963.

Simpson, G. G. *Life of the Past.* New Haven (Yale University Press), 1953.

PALAEONTOLOGY: PLANTS

Andrews, H. N. *Studies in Palaeobotany.* New York (Wiley & Sons), 1964.

PALAEONTOLOGY: VERTEBRATES

Colbert, E. H. *Evolution of the Vertebrates.* New York (Wiley & Sons), 1955.

The Age of Reptiles. New York (W. W. Norton), 1965.

Howell, F. C. *Early Man.* New York (Time-Life Inc.), 1968.

Le Gros Clark, W. E. *History of the Primates.* London (British Museum of Natural History), 5th edn, 1965.

The Fossil Evidence of Human Evolution. Chicago (University of Chicago Press), 2nd edn, 1964.

SUGGESTIONS FOR FURTHER READING

Oakley, K. P. *Man, the Tool-Maker*. London (British Museum of Natural History), 1949.

Romer, A. S. *Vertebrate Palaeontology*. Chicago (University of Chicago Press), 3rd edn, 1966.

Simpson, G. G. *Horses*. New York (Oxford University Press), 1951.

Swinton, W. E. *Fossil Amphibians and Reptiles*. London (British Museum of Natural History), 1958.

Swinton, W. E. *Fossil Birds*. London (British Museum of Natural History), 1958.

Weiner, J. A. *The Piltdown Forgery*. London (Oxford University Press), 1955.

REGIONAL GUIDES AND MAPS

Geological Survey of Great Britain. '*Ten Mile Map*' Sheets *1 and 2*, 1948.

Stose, G. W. *Geological Map of N. America*. (Geol. Society of America), 1946.

Geol. Survey & Museum. *British Regional Geology*. A series of local geological guides covering the whole of Britain. Similar guides are published by some provincial and state surveys in North America.

Trueman, A. E. *Geology and Scenery in England and Wales*. Harmondsworth (Penguin Books), rev. edn, 1971.

Some state surveys in the United States have published guides to the common fossils of their respective areas. Examples are:

Collinson, C. W. *Guides for Beginning Fossil Hunters*. Illinois State Geol. Survey, 1956.

La Rocque, A., and Marple, M. F. *Ohio Fossils*. Columbus, Ohio (Division of Geol. Survey), 1955.

Unklesbay, A. G. *The Common Fossils of Missouri*. Columbia, Missouri (University of Missouri), 1955.

GENERAL REFERENCES

Abercrombie, M., Hickman, C. J., and Johnson, M. L. *A Dictionary of Biology*. Harmondsworth (Penguin Books), 6th edn, 1973.

GLOSSARY

A fuller reference list is contained in A Dictionary of Biology *by Abercrombie, Hickman, and Johnson (Penguin Books), from which some of the present definitions are taken.*

Acanthodians. Extinct group of Palaeozoic fish, often known as 'spiny sharks' but classified as Placoderms. Many of them bore a series of paired spiny fins along the ventral part of the body. Upper Silurian–Permian.

Acclimatization. The responsive adjustment of an organism to a new or changed environment. The change is non-genetic, and may take various forms (physiological, morphological). For example, the greater development of leaves and branches on the sunny side of a tree near the forest margin.

Acraniata. Group of lowly marine chordates, lacking brain, skull, and bony skeletons (cf. the vertebrates or craniates). Includes such forms as *Amphioxus* and graptolites.

Adaptation. Organic characteristics favoured by a particular environment, which may be inherited (cf. the non-genetic acclimatization). Adaptation may also be thought of as the result of natural selection.

Agnatha. Class of vertebrates, represented by living lampreys and hagfish and including the earliest fossil vertebrates (ostracoderms, anaspids, and coelolepids). They lack jaws and paired fins. Ordovician–Recent.

Algae. Sub-division of the Thallophyta. Primitive plants including the seaweeds and diatoms. Pre-Cambrian–Recent.

Amblypods. Group of large, early Tertiary ungulates (including uintatheres and pantodonts). Paleocene–Oligocene.

Ambulacra. The radial areas of echinoderms, along which run the principal nerves and the water tubes.

Ammonites. Group of extinct fossil cephalopod molluscs, related to the living *Nautilus*. Upper Silurian–Upper Cretaceous.

Amphineura. Class of the Phylum Mollusca, including the chitons. A small group of marine animals with some primitive features. Ordovician–Recent.

Angiosperms. The flowering plants, a division of the Tracheophyta. Distinguished from the gymnosperms by having the

GLOSSARY

seeds protected within a closed cavity, the ovary. Triassic?–Recent.

Annelida. Phylum which includes ringed or segmented worms (bristle worms, earthworms, leeches). Pre-Cambrian–Recent.

Archaeocyathids. Extinct (Cambrian) group of sponge-like marine organisms.

Arthrodires. 'Jointed-neck' fishes of Devonian age. A group of the Placoderms.

Arthrophyta. Group of 'jointed plants' belonging to the Tracheophyta, and including the living horsetails (*Equisetum*). Also known as sphenopsids. Devonian–Recent.

Arthropoda. Phylum (the largest) of animals including insects, crabs, ostracods, spiders, centipedes, and the extinct eurypterids and trilobites. Cambrian–Recent.

Asteroidea. Class of Echinodermata, including starfish. Ordovician–Recent.

Aves. The birds. A class of vertebrates: characterized by feathers, warm blood, wings developed from fore-limbs. Jurassic–Recent.

Bacteria. Ubiquitous, unicellular, microscopic organisms, lacking chlorophyll. Probably members of the Thallophyta. Pre-Cambrian?–Recent.

Belemnites. Extinct group of dominantly Mesozoic cephalopod molluscs. The internal fossil shells are cigar-like in shape. Carboniferous–Eocene (or Cretaceous).

Benthic. Organisms living on the sea bottom.

Biocoenose. Well-established community of living organisms, which is in equilibrium with its environment (cf. thanatocoenose).

Biosphere. The total zone of life in, on, and above the earth. It has expanded throughout geological time.

Blastoidea. Class of the Echinodermata. An extinct group of stalked Palaeozoic animals. Ordovician–Permian.

Brachiopoda. Phylum of marine animals, with a two-valved shell: 'lamp shells'. Represented by comparatively few living members, but many fossil forms. Pre-Cambrian?–Recent.

Bryophyta. Division of the Plant Kingdom including liverworts and mosses. A small and rather primitive group of land plants. Carboniferous–Recent.

Bryozoa. Phylum of small, aquatic, usually fixed and colonial

GLOSSARY

animals. Sea mats and 'corallines' (superficially resembling hydroids, but considerably more complex). Often called Polyzoa. Ordovician–Recent.

Caecilians. Order (*Gymnophiona* or *Apoda*) of limbless, burrowing tropical amphibians.

Cambrian. Period of geological time which ended about 500 million years ago. The oldest system of the Palaeozoic era. The oldest group of rocks to contain abundant fossils.

Carboniferous. Period of geological time which ended about 270 million years ago, containing most of the world's major coal-deposits. Often divided into lower (Mississippian) and upper (Pennsylvanian) systems.

Cenozoic. The most recent geological era. The era of recent life, including the Tertiary and Quaternary periods. The 'age of mammals'. The last 70 million years of geological time.

Cephalopoda. Class of Mollusca, including octopus, squids, cuttlefish, *Nautilus*, and the extinct belemnites and ammonites. The head is well developed and bears a crown of tentacles. Cambrian–Recent.

Ceratopsia. The horned dinosaurs, a sub-order of ornithischian dinosaurs. Includes *Triceratops* and other genera of Cretaceous age.

Cetacea. Whales, porpoises, and dolphins. An aquatic order of placental mammals. Eocene–Recent.

Charophytes. The female sex organs ('fruit') of a family (Characeae) of green algae. Represented as fossils by microscopic, striated spheres.

Chlorophyll. Green pigment found in all plants except fungi, bacteria, and a few flowering plants. By means of it plants build up carbohydrates from carbon dioxide and water, absorbing energy from sunlight (photosynthesis).

Choanichthyes. Division of the bony fish (*Osteichthyes*) containing *Crossopterygii* (mainly fossil) and *Dipnoi* (lungfish). Characterized by nasal opening from face to mouth. Devonian–Recent.

Chondrichthyes. Class of vertebrates, containing the cartilaginous fish: sharks, skates, rays, chimaeras. Devonian–Recent.

Chordata. Phylum of animals with a notochord (q.v.), hollow dorsal nerve cord, and gill slits. Includes vertebrates and Acraniata (q.v.). Cambrian–Recent.

Coelenterata. Phylum of animals including hydroids, jellyfish,

GLOSSARY

sea anemones, and corals. All aquatic, mostly marine. Simple body structure. Pre-Cambrian?–Recent.

Commensalism. Form of symbiosis (q.v.) in which only one of the partners derives benefit from the association, but in which the other is unharmed.

Community. Group of organisms of one or more species living together and representing a closely interacting system. A group of populations inhabiting a given area.

Conifers. Group of gymnosperm (q.v.) plants. Includes pine, spruce, cedar, yew, larch, etc. Usually tall evergreen forest trees.

Conodonts. Group of extinct, dominantly Palaeozoic tooth-like fossils of obscure affinities. Cambrian–Triassic.

Cordaites. Extinct group of conifer-like trees, belonging to the gymnosperms. Largely confined to the Upper Palaeozoic.

Cotylosaurs. The 'stem reptiles'. Primitive reptiles of Upper Carboniferous–Triassic age, from which many later forms developed.

Craniata. The vertebrates, having a skull and bony skeleton (cf. the Acraniata). Ordovician–Recent.

Creodonts. Primitive carnivores, archaic in structure, generally small and slender. Early Tertiary.

Cretaceous. Period of geological time which ended about 70 million years ago. The last system of the Mesozoic era.

Crinoidea. Class of Echinodermata, including feather stars and sea lilies. Usually fixed by a long stalk, but some forms are free-swimming. Ordovician–Recent.

Crossopterygii. Group of bony fish. Sub-order of the Choanichthyes (q.v.) 'Coelacanths', represented by only one living genus (*Latimeria*). Fossil forms represent the ancestors of the land vertebrates. Devonian–Recent.

Cuticle. Superficial non-cellular layer covering an animal or plant. In higher plants forms a continuous protective layer, and prevents excessive moisture-loss.

Cycadeoids. Group of extinct Mesozoic gymnosperm (q.v.) plants, differing from living cycads in their method of reproduction.

Cycads. Group of gymnosperms, which are the most primitive living seed plants. Permian–Recent.

Cystoidea. Extinct class of Palaeozoic Echinodermata. 'Bladder-like' in form, usually attached. Ordovician–Devonian.

GLOSSARY

Cytoplasm. All the protoplasm of a cell excluding the nucleus.

Devonian. Period of geological time, which ended about 350 million years ago. Often spoken of as the 'age of fishes'.

Diatoms. Group of Algae. Microscopic, unicellular plants, solitary or colonial. Siliceous 'skeletons' form diatomaceous earth deposits.

Dicotyledon. One of two groups of flowering plants, having two seed leaves, net veined leaves, etc. Includes most forest trees, fruits, vegetables, and flowers. Jurassic?–Recent.

Dipnoi. The lungfish. A group of Choanichthyes (q.v.). Represented by three living fresh-water genera. Devonian–Recent.

Echinodermata. Phylum of animals, including sea urchins, sea cucumbers, brittle stars, starfish, sea lilies, and feather stars. Marine, usually radially symmetrical, with water vascular system. Include extinct fossil blastoids, cystoids, eocrinoids, edrioasteroids (q.v.). Cambrian–Recent.

Echinoidea. Sea urchins. Class of Phylum Echinodermata (q.v.). Usually globular, spiny skeleton. Ordovician–Recent.

Ecology. The study of the relationship between organisms and their environment.

Edrioasteroidea. Extinct class of Palaeozoic Echinodermata. Globular body containing five sinuous arms. Cambrian–Lower Carboniferous.

Eleutherozoa. Group of free-living Echinodermata. Includes sea urchins, sea cucumbers, brittle stars, and starfish (cf. Pelmatozoa).

Embolomeres. Group of labyrinthodont amphibia. The most primitive labyrinthodonts. Devonian–Permian.

Eocene. Sub-division of the Tertiary system. Ended about 40 million years ago.

Eocrinoidea. Extinct group of Echinodermata. Primitive attached Lower Palaeozoic crinoid-like creatures. Cambrian–Ordovician.

Eucaryotic cells. Nucleus enclosed in a membrane, and capable of mitotic cell division.

Eurypterids. Extinct group of Arthropoda. Palaeozoic. Aquatic scorpions, up to nine feet in length. Ordovician–Permian, but especially characteristic of Silurian and Devonian.

Fermentation. Metabolism in which sugars are converted to

GLOSSARY

alcohols, without the involvement of oxygen, but with release of energy.

Filicineae. Group of tracheophyte plants: ferns. Devonian–Recent.

Fissipedes. The 'split-footed' carnivores. Mammalia, including dogs, bears, raccoons, otters, weasels, hyenas, and cats. The dominant carnivores (cf. Pinnipedes). Paleocene–Recent.

Flagellates. Class of Phylum Protozoa whose members have one or more long thread-like 'tails' (flagella). Includes both plant- and animal-like forms, and some with mixed characters.

Foraminifera. Group of protozoans, mostly marine, many of which developed shells, usually calcareous. Most are microscopic. Ordovician–Recent.

Fossil. The remains of, or direct indication of the existence of, life of the geological past.

Fungi. Sub-division of Thallophyta. Mushrooms, moulds, yeasts, etc. Simply organized plants, lacking chlorophyll.

Fusulinids. Group of foraminifera whose shells resemble grains of wheat, and have a very complex internal structure. Carboniferous–Permian.

Gastropoda. Class of Mollusca. Includes snails, slugs, and pteropods. Marine, fresh water, and terrestrial. Distinct head. Often a single coiled shell. Cambrian–Recent.

Geosaurs. Group of extinct reptiles, Thallatosuchians. Marine crocodiles. Jurassic.

Geosynclines. Areas of considerable extent (often linear) of more or less continuous subsidence over long periods, which usually receive great thicknesses of sediments and volcanic rocks.

Graptolites. Extinct, colonial, marine organisms, apparently related to the Hemichordates. Cambrian–Carboniferous.

Gymnosperms. Conifers and their allies (Cycads, Ginkgos, Cycadecids, Cordaites). Primitive seed plants, differing from angiosperms (q.v.) by having naked, unprotected seeds. Devonian?–Recent.

Hemichordates. Small group of marine animals, including the acorn-worm *Balanoglossus* and probably the extinct graptolites. Primitive chordates. Cambrian–Recent.

Holothuroidea. Sea cucumbers. Class of Echinodermata. Soft cylindrical body, bearing microscopic 'plates'. Cambrian–Recent.

GLOSSARY

Hyomandibular. The dorsal element (bone or cartilage) of the hyoid arch, which takes part in the jaw suspension of most fish, and in tetrapods becomes one of the ear-bones.

Ichthyostegids. Primitive, extinct amphibia, exhibiting many fish-like characteristics. Devonian–Carboniferous.

Igneous rock. Rock formed by solidification of hot mobile material called magma. Magma originates within the earth, and occurs mostly in a liquid silicate molten phase, various solid phases (such as suspended crystals), and often gas.

Jurassic. Period of geological time which ended about 135 million years ago. The middle system of the Mesozoic.

Labyrinthodonts. Primitive group of squat amphibia. The dominant amphibians of late Palaeozoic and early Mesozoic times. Devonian–Triassic.

Lepidophyta. An alternative name for the Lycopsida, or lycopods (q.v.).

Lycopods. Club mosses (Lycopodiales). Includes giant Coal Measure scale trees (*Lepidodendron, Sigillaria*). Devonian–Recent.

Mammalia. Class of tetrapod vertebrates. Man, dog, whale, etc. Characterized by hair, milk secretion, diaphragm used in respiration. Jurassic–Recent.

Marsupials. Group of Mammalia, from Australia and North and South America. Kangaroo, opossum, etc. Young are born in very undeveloped state and sheltered in mother's pouch (*marsupia*). Cretaceous–Recent.

Mesosaurs. Primitive aquatic reptiles. Upper Carboniferous.

Mesozoic. The middle of the three geological eras of life, including the Triassic, Jurassic, and Cretaceous periods. Began about 225 million years ago and lasted 155 million years – 'the age of reptiles'.

Metamorphic rock. Rock formed by alteration of a pre-existing rock as a result of pronounced changes in temperature, pressure, and chemical environment.

Metamorphism. Changes in the texture or composition of a rock, after its induration or solidification, produced by increased temperature, pressure, or chemical changes.

Miocene. Sub-division of the Tertiary System. Ended about 11 million years ago.

Mollusca. A large Phylum of animals including snails, mussels,

GLOSSARY

cephalopods, etc. Mostly aquatic, soft bodied, with a hard shell, unsegmented, with a head and muscular foot. Cambrian–Recent.

Monocotyledon. Smaller of the two groups of flowering plants (angiosperms). Single seed leaf, parallel veined leaves, flower parts usually in threes or multiples thereof. Includes lily, tulip, orchid. Triassic?–Recent.

Monotremes. Sub-class of mammals. Duck-billed platypus and two genera of spiny anteater. Australia and New Guinea. Very primitive group. Lay eggs and have some reptilian characters. Pleistocene–Recent, but probably much older.

Mosasaurs. An extinct group of marine reptiles: 'Sea lizards'. Cretaceous.

Mutualism. Form of symbiosis (q.v.) in which the association is mutually beneficial and without damage to either partner.

Notochord. Skeletal rod, lying lengthwise, between nerve-cord and gut. Present in some stage of development in all chordates.

Nucleus. Body containing the chromosomes, present in nearly all cells of plants and animals and perhaps in some bacteria: not in viruses.

Oligocene. Sub-division of the Tertiary system. Ended about 25 million years ago.

Ophiuroidea. Brittle stars. Class of Echinodermata. Star-shaped, long sinuous arms radiating from central disc. Ordovician–Recent.

Ordovician. Period of geological time. Part of the Lower Palaeozoic era. Ended about 440 million years ago.

Ornithischia. Fossil order of Reptilia. 'Bird-hipped' dinosaurs, the pelvis resembling that of birds. Triassic–Cretaceous.

Osteichthyes. Class of Chordata. Bony fish, includes *Actinopterygii* and *Choanichthyes* (q.v.). Devonian–Recent.

Ostracods. Group of generally minute, bivalved crustacean arthropods. Cambrian?–Recent.

Palaeozoic. The oldest of the three geological eras of life. The era of ancient life: the 'age of invertebrates'. Includes the Cambrian to Permian systems. Began about 600 million years ago and lasted about 375 million years.

Palaeontology. The study of ancient life: the study of fossils.

Paleocene. The lowest sub-division of the Tertiary system.

GLOSSARY

Ended about 60 million years ago. The 'dawn of the age of mammals'.

Parasitism. Form of symbiosis: an involuntary association in which progressive damage is done to the host, to the benefit of the parasite.

Pelecypoda. Class of Phylum Mollusca Lamellibranchia. Mussels, clams, oysters, etc. Aquatic, bivalved molluscs. Ordovician–Recent.

Pelmatazoa. Group of sedentary Echinodermata. Includes crinoids, cystoids, blastoids, etc. (cf. *Eleutherozoa*).

Pelycosaurs. Group of extinct aberrant mammal-like reptiles. Includes the sail-back lizards and others from which the mammals probably arose. Carboniferous–Permian.

Permian. Period of geological time. The last system of the Palaeozoic era, which ended about 225 million years ago.

Photosynthesis. Synthesis by green plants of carbohydrates from water and carbon dioxide, with the aid of energy absorbed from sunlight (see chlorophyll).

Phytosaurs. Extinct group of thecodont reptiles: 'Plant-lizards', it being formerly and wrongly supposed that they were herbivores: crocodile-like reptiles. Triassic.

Pinnipedes. The feather-feet marine carnivores. Includes seals, sea lions, walruses, and sea elephants. Miocene–Recent (cf. Fissipedes).

Pisces. 'Fish'. A super-class including four classes of fish. (Agnatha, Placoderms, Chondrichthyes, Osteichthyes, q.v.).

Placentals. Mammals belonging to the sub-class Placentalia, containing most living mammals. The embryo develops in the uterus, attached by a highly organized 'placenta'.

Placoderms. Extinct class of vertebrates (sometimes called Aphetohyoidea). Primitive fish, with archaic jaw suspension. Upper Silurian–Permian.

Plankton. Floating or drifting organisms of seas or lakes. Mostly small, existing near the surface, where light is plentiful. Of great importance as food for fish and whales.

Pleistocene. Sub-division of the Quaternary Period of geological time. It ended about 11,000 years ago. An episode of widespread continental glaciation in the Northern Hemisphere.

GLOSSARY

Pliocene. Subdivision of the Tertiary Period of geological time. The most recent division of the Tertiary. Ended about 1 million years ago.

Population. Single or mixed species association that presents a closely interacting system (in food competition, etc.).

Porifera. Sponges: phylum of multicellular but very primitive animals. No nervous system. 'Collar' cells. Often develop spicules. Cambrian–Recent.

Pre-Cambrian. Vast interval of geological time, preceding the time of deposition of the oldest fossiliferous rocks (the Cambrian). The first 4,000 million or so years of earth history.

Procaryotic cells. Cells without a nuclear wall, which do not undergo mitotic cell division.

Protozoa. Phylum of animals, differing from all others (Metazoa) in consisting of one cell only. Includes some plant-like forms, and regarded by some as a third kingdom (Protista). Includes foraminifera and radiolaria. Ordovician–Recent, but almost certainly older.

Psilophyta (Psilopsids). Group of very primitive vascular plants (tracheophytes). The first terrestrial plants. Silurian–Recent.

Psilopsids. See *Psilophyta*.

Pteridophytes. Ferns, horsetails, club mosses, etc. Division of plant kingdom, including spore-bearing terrestrial plants. Well-developed sporophyte, alternation of generations, well-developed roots, stem, and leaves. Includes Filicales (ferns), Arthrophyta, Lepidophyta, Psilophyta. Silurian–Recent.

Pterobranchs. Group of hemichordates, having notochord during part of their life history. Probably includes the extinct graptolites and the living genera *Cephalodiscus* and *Rhabdopleura*.

Pterodactyls. See *Pterosaurs*.

Pteropods. Sea butterflies. Gastropod molluscs, highly modified for pelagic life, with the foot extended to form 'wings'. Cambrian?–Recent.

Pterosaurs (Pterodactyls). Fossil order of flying reptiles. Wing membranous, supported by greatly elongated fourth finger. Triassic?–Jurassic–Cretaceous.

Quaternary. The most recent period of geological time. The last 1,000,000 years or so. Includes the Pleistocene and the Recent epochs.

GLOSSARY

Radiolaria. Group of marine planktonic protozoans, having delicate siliceous skeletons. Pre-Cambrian?–Recent.

Reptiles. Class of vertebrates. Includes turtles, lizards, snakes, crocodiles, and many extinct forms, such as dinosaurs. Cold-blooded tetrapods, dominant in the Mesozoic: amniote egg. Carboniferous–Recent.

Rhachitomes. Group of fossil amphibia. The most advanced members of the labyrinthodonts. Carboniferous–Permian.

Saltation. Apparent evolutionary change by abrupt transition, without intermediate ancestors.

Saurischia. Order of fossil reptiles, to which many of the dinosaurs are referred (the other is Ornithischia, q.v.). 'Lizard-hip' structure. Triassic–Cretaceous.

Sauropods. Group of generally large (up to fifty tons) quadrupedal, herbivorous dinosaurs. Triassic–Cretaceous.

Scaphopoda. Elephant tusk shells. Small class of burrowing marine molluscs, with tubular shell. Silurian–Recent.

Sedimentary rocks. Rocks formed either from sediment composed of fragments of earlier rocks (e.g. sandstone), or by precipitation from solution (e.g. rock salt), or by secretions of various organisms (e.g. some limestone).

Silurian. A period of geological time, which ended about 400 million years ago. Part of the Lower Palaeozoic era.

Sirenia. Manatees, dugongs, sea cows. An aquatic order of placental mammals, having front flippers and vestigial hind legs. Related most closely to elephants and conies. Paleocene–Recent.

Species. Naturally interbreeding populations, which are reproductively isolated from other such groups. The smallest unit of classification commonly used.

Sphenopsids. See *Arthrophyta.*

Sporangium. The organ of a plant, within which are produced asexual spores.

Spores. Reproductive bodies, usually minute, that become detached from the parent and produce new individuals. Found in all groups of plants and in protozoans.

Stegocephalia. See *Labyrinthodonts.*

Stereospondyls. Group of 'degenerate' labyrinthodont amphibia. Triassic.

Symbiosis. The mutual intimate association of specific plants with

GLOSSARY

plants, plants with animals, or animals with animals (see commensalism, mutualism, parasitism).

Synapsids. Large group of reptiles, having a lower temporal opening behind the eye. Includes pelycosaurs and therapsids. Carboniferous–Triassic.

Taxonomy. The science of the classification of living things.

Test. A shell, hard covering, or exoskeleton.

Tetrapods. Four-footed animals. A group of vertebrates. Includes amphibia, reptiles, birds, and mammals.

Thallophyta. Division of plant kingdom, containing the most primitive forms of plant life. Includes algae, fungi, and (probably) bacteria. Pre-Cambrian–Recent.

Thanatocoenose. A 'fossil community'. A death assemblage of organisms preserved in a rock, differing from the biocoenose (q.v.) in the absence of some original members and in the addition (after death) of some exotic forms.

Thecae. The minute cap-like structures on graptolite fossils in which individual organisms were housed.

Thecodonts. Group of reptiles in which the teeth are set in sockets in the jaws. 'Stem reptiles'. Includes bipedal forms and crocodile-like forms. Triassic.

Therapsids. Group of reptiles with many mammal-like features. Includes the ancestors of the mammals (ictidosaurs); also dinocephalids, dicynodonts, and theriodonts. Permian–Triassic.

Theropods. 'Beast-footed' dinosaurs. A group of the saurischians including bipedal carnivorous dinosaurs. Triassic–Cretaceous.

Tracheophyta. The largest and most advanced division of the plant kingdom. The vascular plants. Silurian–Recent.

Triassic. Period of geologic time. The oldest period of the Mesozoic era. Ended about 180 million years ago.

Trilobites. Extinct group of arthropods: marine, widespread in Lower Palaeozoic. Cambrian–Permian.

Tube-feet (Podia). Hollow, extensile appendages of Echinodermata: connected to water-vascular system (q.v.). In some groups (starfish, sea urchins) they end in suckers and assist in locomotion; in others (brittle stars, crinoids) suckers absent.

Tunicates (Urochordata). Sea squirts, etc. A group of primitive marine chordates. Adult sedentary, gill slits, reduced nervous system, no notochord; larva active, tadpole-like, well-developed nervous system and notochord.

GLOSSARY

Uintatheres. Group of archaic mammals, with horn-like protuberances, and of heavy build. Paleocene–Oligocene.

Ungulates. Hoofed mammals. Usually adapted for running on firm, open ground. Herbivorous, living in herds. (See Artiodactyla and Perissodactyla.)

Vascular tissue. Plant tissue forming a continuous system throughout higher plants. Functions in conduction of water, mineral salts, and synthesized food materials, and in mechanical support.

Virus. Variable group of disease-producing agents, parasitic in plants and animals, unable to multiply outside the host tissues, of very problematical affinities.

Water vascular system. Of echinoderms. System of canals containing fluid, open to either sea or body cavity, which supplies water to the tube-feet (q.v.).

INDEX

Absolute time, determination of, 72–5
Acanthodians, 167
Acclimatization, 31
Acorn-worms, 161
Acraniata, 158
Actinopterygians, 172
Adam, 21
Adaptation, 31
Age of the earth, 73, 75–6
Ages, Dark, 35; Middle, 35
Agnatha, 163–5
Air, conquest of, by arthropods, 140, 211–13; by birds, 217–19; by mammals, 220; by reptiles, 208, 213–16
Albertus Magnus, 35
Algae, 26, 98, 145
Allosaurus, 199
Amblypods, 235–6
Ammonites, 131–2, 274, 299
Amniote egg, 185–6
Amoeba, 125–6
Amphibia, 177; evolution from fishes, 176–7, 294; history of, 180–84; life of, 177–8
Amphineura, 105
Amphioxus, 161
Angiosperms, characters of, 224; distribution of, 224; history of, 223–5; significance of, in evolution of mammals, 225–6
Animals, major groups of, 26, 28
Ankylosaurus, 203
Annelids, 28, 107–9
Annularia, 152, 154
Anteaters, 230, 249
Apes, 252; anthropoid, genera of, 254; similarities to man, 254
Archaeopteryx, 55, 56–7, 218, 295
Archimedes, 130

Aristotle, 23, 24, 35
Armadillos, 248–9
Arthrodires, 167
Arthropods, 28; fresh-water, 140–42; groups of, 115–18; history of, 115–18; reasons for success in air of, 140–42
Artiodactyls, 239; distribution of, 245; history of, 245
Asteroidea, 132
Augustus, 34
Australopithecus, 255, 257–8
Avicenna, 35

Bacteria, 19, 26, 145–9
Balanoglossus, 160–62
Baluchitherium, 244
Barnacles, 115, 274
Bats, 220
Bauer, G., 36
Becquerel, 74
Belemnites, 58, 59, 131, 272
Bennettiales, 221
Beringer, J. B., 36
Biochemistry, 302, 303
Biophysics, 302
Biosphere, 29
Birds, characters of, 57, 217; history of, 217–19
Birkenia, 165
Biston betularia, 286
Blastoidea, 109, 134–5, 299
Borings, 42
Brachiopods, 28, 58, 112; articulate, 113–15, 124; history of, 112–13; inarticulate, 112–15
Brain, evolution of, 242
Brontosaurus, 199
Broom, R., 257
Bryophytes, 20, 28; history of, 149

323

INDEX

Bryozoa, 28, 129–30
Buffon, 284
Burrows, 42

Cain & Sheppard, 286
Cambrian, 71; diversity of fossils of Lower, 77–81, 90; faunal provinces of, 119; invertebrates of, 119–20; life of, 78–81, 279
Captorhinomorphs, 191
Carboniferous, 71; conditions of, 138–9, 156–7; life of, 141–3, 151–6
Carnivores, 234–7
Casts, 42
Category, taxonomic, 21
Cats, 237
Cave art, 266–7
Cave bears, European, 287–90
Caytoniales, 223
Cell, 17, 27
Cellular structure, 302
Cenozoic, 69, 71; age of mammals, 234; geography of, 234; life of, 227–8, 229–51
Cephalaspids, 163–6
Cephalopods, 131–2, 274
Ceratopsians, 203
Cetaceans, 250
Characters, 25
Chitons, 105–6, 301
Chlorophyll, 26
Choanichthyes, development of lungs in, 174
Chondrichthyes, 169–71
Chordates, 28, 158; primitive, 160
Chromosomes, 301
Cladoselache, 171
Class, 24, 25
Classification, 23–4
Coal Measures, climate of, 152, 156; forests of, 152–3; geography of, 156; life of, 152–6

Coal swamps, 151
Coelacanth, 175, 298
Coelenterates, 28, 103–4
Commensalism, 31–2
Communities, 31
Condylarths, 235
Coniferophyta, 155
Conifers, 155, 221–3, 225–6
Convergence, 198, 205, 250–51, 281
Convergent evolution, in South American mammals, 247
Copernicus, 277
Coprolites, 42
Coral reefs, fossil, 128
Corals, 127–9, 270
Cordaites, 155–6
Correlation, geophysical methods of, 66; methods of, 65–7; use of fossil studies in, 58, 61, 66, 68
Cotylosaurs, 193
Crabs, 115, 274
Creodonts, 235
Cretaceous, 71; changed conditions leading to life changes at close of, 209–10; extinction of reptiles at close of, 208–10, 234; life of, 197–208, 221–7, 229–33, 298, 299
Crinoids, 112, 134–5, 274
Crocodiles, 197; history of, 203–4
Crossopterygians, 174, 294; affinities of, 175–7; evolution of higher vertebrates from, 177–80
Crustaceans, 117, 135
Cryptozoic Era, 69
Cuvier, 37
Cycadeoids, 221–3, 226
Cycads, 221–2, 226
Cynognathus, 196–7
Cystoids, 112, 134
Cytoplasm, 26

Da Vinci, Leonardo, 36
Dart, R., 257

324

INDEX

Darwin, Charles, 19, 37, 79, 109, 255, 277–8, 283–5, 290–91, 296, 301, 302–3
Darwin, Erasmus, 284
Dating, 54
Davies, Morley, 297
Dawson, Charles, 53–4
De Beer, Sir Gavin, 56, 57, 187, 295
Devonian, 71; conditions of, 121; life of, 122–38, 148, 164, 166, 169, 170, 298, 299
Diadectomorphs, 191
Diatoms, 147
Dibranchiates, 272
Dicotyledons, 224
Dicynodonts, 196
Dimetrodon, 195
Dinichthys, 167
Dinocephalians, 196
Dinosaurs, 197–203
Diplodocus, 199
Dipnoi, 282
Dogs, 237
Dolphins, an example of convergent evolution, 250
Drosophila, 302
Dubois, E., 259
Duck-billed platypus, 230

Echinoderms, 28, 109–12, 132, 274; possible ancestors of vertebrates, 162–3
Echinoidea, 133–5, 274
Ecology, 29
Edaphosaurus, 195
Edentates, 248
Edrioasteroidea, 111, 134–5
Egg, amniote, 185–6
Electron microscope, 26, 302
Elephants, history of, 245–6
Eleutherozoa, 133
Elpistostege, 179
Embolomeres, 181
Emerson, 15

Environment, importance of, in evolution of life, 29, 279
Eocene, 70, 232, 237, 292
Eocrinoids, 112, 134
Epiceratodus, 174
Epstein, 58
Eryops, 180, 183
Euglena, 144–5
Eurypterids, 115, 117, 136, 140
Eusthenopteron, 175, 178
Evolution, importance of geographical isolation in, 231, 246–9; mechanism of, 283–305; mosaic, 56, 187; process of, 280; rates of, 282–3, 304–5
Expansion of life, periods of rapid, 120, 279
Extinction, periods of, 120, 169, 189–90, 208, 279; causes of, 209–10

Family, 25
Faunas, 48
Felicineae, 153
Ferns, 154, 221
Fish, history of, 163, 173; orders of, 163–77; primitive, 163
Fissipedes, 237
Flagellates, 26
Flora, Angaran, 156; Cathaysian, 156; changing distribution of, in Cenozoic, 227–8; *Dicroidium*, 221; Gondwanan, 156
Floral migration in Pleistocene, 227–8
Floral zones in Cenozoic, 227–8
Foraminifera, 125–7, 293
Fossil, assemblages, study of, 48; environment, study of, 52; record, biggest gaps in, 88–92; studies, value of, 60, 62
Fossilization, process of, 38
Fossils, assemblages of, 48–9, 67; collection of, 45–7; distribution of, 51; earliest, diversity

INDEX

Fossils – *contd*
of, 77; nature of, 34; occurrence of, 43; preparation of, 45, 47; study of, 46, 50–52, 278–9, 290–301
Frogs, 178, 184
Functional data, study of, 50–51
Fungi, 144–6
Fusulinids, 139

Gastropods, 104–7, 130–31, 272, 301; earliest fresh-water, 142; earliest land, 142
Genes, 301–2
Genetics, 302
Genus, 24, 25, 281
Geological age, of the earth, 75; use of radioactive methods in determination of, 74
Geological time, 64; divisions of, 67–71; scale, 65, 69–72, 74, 76
Geosaurs, 204, 207
Ginkgos, 223
Glossopteris, 156
Glyptodonts, 249
Gondwanaland, 138
Goniatites, 132
Graptolites, 137–8
Grasses, history of, 228; significance of, in evolution of the mammals, 228
Gymnosperms, 224

Heredity, 302
Herodotus, 35, 72
Hesperornis, 218
Hierarchy, taxonomic, 24
Holosteans, 275
Holothuroids, 112
Homo erectus, 259–61
Homogalax, 296
Homo heidelbergensis, 261
Homo neanderthalis, 262–3; belief in survival of, 263; extinction of, 263; mode of life of, 263; tools of, 262

Homo sapiens, 254, 260, 263
Hoofed placentals, 234–5
Hooke, Robert, 37
Horse, history of, 238–9, 243, 296, 297, 298; significance of evolutionary changes in, 238–43, 292
Horsetails, 152
Human cultures, 255
Huxley, 292
Hydra, 103
Hyracotherium, 292

Ice Age, 255, 260
Ichthyornis, 218
Ichthyosaurs, 206–7
Ichthyostegids, 178–80, 294
Ictidosaurs, 196–7, 229
Igneous rocks, 44
Iguanodon, 201
Implements, as indicative of human remains, 43, 260–61, 262–3
Insectivores, 231, 252
Insects, 117, 140, 211–13; history of, 211–12; life of, 212–13
Invertebrates, fossils, 98–143; kinds of, 28
Isolation, role of, in evolution, 249
Isotope studies, 58, 74

Jamoytius, 163
Jellyfish, 103–4
Jurassic, 71; conditions of, 189–90, 223; life of, 61, 189–90, 197–208, 221–3, 229–33, 299

Kelvin, Lord, 73–4, 304
Kennett, 293
Kettlewell, H. B. D., 285–6
Kingdom, animal, 24, 28; plant, 24, 28
Knight, 301
Kovalevsky, 292
Kurten, 287–90

INDEX

Labyrinthodonts, 180; characters of, 180; history of, 180–84
Lakes and rivers, conquest of, 139–43
Lalicker, C. G., 60
Lamarck, J. C. de, 37, 284
Land, conquest of, by amphibia, 177–80; by arthropods, 140; by mammals, 231; by reptiles, 184–6
Le Gros Clark, Sir Wilfred, 54
Lemurs, 252
Lepidodendron, 153, 154
Lepidophyta, 152
Lhuyd, E., 36
Linnean Society, 277
Life, abundance of, 18–19, 285; continuity of, 279–80, 291–4; distribution of, 29, 32; diversity of, 20, 291; emergence of, 77–9; environment of, 280–81; evolution of, 280; expansion of, 279; extinction of, 280–81; factors controlling the distribution of, 29; history of, 70; interrelationships of, 280; nature of, 17; origin of, 92–7; persistence of, 281–2; and time, 69–72
Limulus, 136
Lingula, 112–15, 283
Linnaeus, Carl von, 24, 25
Lister, Martin, 37
Littleton, R. A., 75
Lizards, history of, 204–5, 207
Lobsters, 115, 274
Lowenstam, H. A., 58
Lungfish, 174, 282
Lycopods, 150, 152, 154

Mammals, characters of, 229; flying, 220; history of, 229–33, 280; marine, 250–51; orders of, 229–33; origin of, 229; South American, 230, 246–9

Man, ancient, 253–62; Cro-Magnon, 255, 265–7; burial of, 265; cave paintings of, 267; definition of, 254; descent of, 255; history of, 257; mode of life of, 267; tools of, 263–7; modern, 263–4; Neolithic, and his use of tools, 261, 267; Piltdown, 53–5, 257; Swanscombe, 263; tool-maker, 261–2
Marsh, 292
Marsupials, characters of, 230; history of, 230; orders of, 230–31
Mayr, E., 20, 23
Megatherium, 249
Melanism, 285–6
Mendel, Gregor, 301
Mesosaurs, 193
Mesozoic, 69; Age of the Reptiles, 197; conditions of, 225; life of, 197, 221, 269–76, 279, 298
Metabolic processes, 302
Metamorphic rocks, 44
Microfossils, uses of, in correlation, 61
Microsaurs, 184
Miller, S., 95
Millipede-like creatures in Upper Silurian, 140
Miocene, 70, 232, 237, 241, 243, 293
Missing links, 55, 56, 179
Mollusca, 28, 104–7, 142; divisions of, 105–7; history of, 104–7, 130, 270–75, 298, 301
Monkeys, distribution of, 249, 252–4; history of, 249, 252–4
Monocotyledons, 224
Monotremes, 197, 230
Mosasaurs, 204, 207
Moths, 285–6
Moulds, 42
Mousterian culture, ceremonial burial in, 261

327

INDEX

Multituberculates, 230
Mutation, 27
Myzophyceae, 26, 56

Natural selection, criticisms of, 53, 56, 79; evidence in favour of, 285–90, 304; role of, in evolution, 285–90; time required for, 304–5
Nautiloids, 132, 274
Nautilus, 131
Neanderthal Man, 256, 262–3
Neolithic Age, 267
Neopilina galathea, 298, 301
Neotrigonia, 298
New Zealand, 293
Newton, 277
Nomenclature, 23
Nothosaurs, 207
Notochord, 158
Nucleic acids, 17, 26, 302
Nucleus, 26

Oakley, K. P., 54
Octopus, 131
Oligocene, 70, 232, 237
Oparin, A., 94
Ophiuroidea, 133
Order, 24, 25
Ordovician, 71; conditions of, 122; life of, 120–38
Origin of earth, 75
Origin of life, mechanisms of, 92–7
Origin of Species, 37, 278, 284
Ornamentation, as an adaptation to mode of life, 123–4
Ornithischia, 198, 201
Ornithopods, 201–2
Osteichthyes, 171
Ostracoderms, oldest fossil vertebrates, 163–6
Ostracods, 135–6
Outcrop, 47, 64
Owen, Sir Richard, 55, 198, 296

Palaeolithic Man, 261–7

Palaeoniscids, 275
Palaeontology, 38
Palaeozoic, 69; Lower, conditions of, 99; life of, 98–120, 279; Middle, climate of, 121; geography of, 121; life of, 121, 149, 164, 170; Upper, climatic changes at close of, 138–9, 189; extinction of life at close of, 169, 189–90; glaciation, 139, 189–90; life of, 139, 149, 154, 188, 298
Paleocene, 70, 232, 235
Pantotheres, 230
Paradoxides, 117
Parasitism, 32
Pareiasaurs, 193
Pelecypods, 58, 105, 106, 130–31, 272
Pelmatozoa, 133
Pelycosaurs, 193–5
Pentacrinus, 274
Perissodactyls, 239
Permian, 71; conditions of, 138–9; geography of, 138; life of, 139, 191–8, 299
Permineralization, 41
Phylogeny, 26
Phylum, 25, 281
Phytosaurs, 197
Piltdown Man, 53–5
Pisaster, 19
Pisces, 159
Pithecanthropus, 255, 259, 295
Placentals, 231–3
Placoderms, 165–9
Placodonts, 207
Plants: Cenozoic, 227; differences between, and animals, 144–5; kinds of, 28, 144; Mesozoic, 221–7; primitive land, in Upper Silurian, 139, 150; Palaeozoic, 150; role of, in evolution of arthropods and vertebrates, 151

INDEX

Pleistocene, climate of, 228, 255, 260; life of, 255, 260; man in, 255, 261
Plesiosaurs, 207–8
Pliocene, 70, 232; life of, 232, 293; man in, 257
Population, 31
Porifera, 28, 99–103
Prairie, spread of, 228
Pre-Cambrian, 72; evidence for life in, 77–81, 83–8, 304; fossil gap between, and Lower Cambrian, 76, 77, 87, 89–92; rocks of, 82–3
Preservation of hard parts, 41; of soft parts, 40
Primates, characters of, 252–68
Proteins, 17, 26
Protista, 24
Protoceratops, eggs found in Mongolia with, 42, 203
Protoplasm, 17
Protorosaur, 213
Protosuchus, 204
Protozoa, 28, 124–7
Psilophyta, 151
Pteranodon, 215–16
Pteridospermaphyta, 155
Pterodactyls, 208
Pterosaurs, 213; characters of, 213; history of, 213–16

Quaternary, 71, 232

Radiolaria, 125–7
Ramapithecus, 257–8
Ray, John, 24
Rays, 275
Realm, zoogeographic, 32
Reptiles, 184; decline of, 203, 208, 227; differences between amphibia and, 183, 185–7, 191; explosive evolution of, as adaptive radiation, 193; flying, 213–16; history of, 192, 194, 300; life of, 185–6; marine, 199, 205–8, 275, 281; reasons for success on land of, 185–6, 191; reproduction of, 185
Rhachitomes, 180
Rhinoceros, 244
Rhynchocephalians, 204
Rodents, 246
Romer & Whittier, 300
Rudistids, 272
Rugose corals, 128

Saurischia, 198
Sauropods, 199
Scaphopoda, 105
Scheuchzer, Johann, 36
Scolecodonts, 107
Scorpions, oldest, 117, 136, 140
Sea lions, 251
Seals, 251
Seas, life in Palaeozoic, 98–139; Mesozoic, 270–75; Cenozoic, 270–75
Sediments, fossils of, 44–5; deposition of, 44
Seed, origin of, 153; significance of, 153, 303
Seymouria, 183, 186–7
Seymouriamorphs, 186, 295
Sharks, history of, 169–71, 275
Sigillaria, 153
Silliman lectures, 300
Silurian, 71; climate of, 121; life of, 120–38
Simpson, G. G., 90, 95, 296, 298, 299
Sirenia, 251
Skates, 275
Sloths, 249
Smith, William, 37, 66–7
Snails, 286–7
Snakes, history of, 204–5
Solenhofen, first fossil bird at, 55; Jurassic rocks of, 55
South America, extinction of mammals in, 230, 246–9;

329

INDEX

South America – *contd*
  orders of mammals of, 248–9
Species, 25; number of, 20
Sphenodon, 204
Sphenopsids, 152
Sponges, 99–103
Stegosaurs, 201
Stegosaurus, 200
Stereospondyls, 182
Stone Age, climate of, 260–61; man in, 260–61
Stromatoporoids, 129
Symbiosis, 31
Symmetrodonts, 230
Synapsids, 195
Systema Naturae, 24

Tapirs, 244, 296
Tarsioids, 252
Taxonomist, 21, 26
Teeth, 240–41, 243
Teleosts, trends in development of, 275
Temperature, studies of, 58, 59
Tertiary, 71, 227–8, 229–33, 234–51, 280, 293
Tetrabranchiates, 274
Tetrapods, 177
Textularia miozea-kapitea, 293
Thallophytes, 20, 28, 144–9
Thecodonts, 213; characters of, 197; emergence of dinosaurs from, 198
Theophrastus, 35
Therapsids, 196
Theriodonts, 196, 229
Time, absolute, 72–5
Titanotheres, 244–5
Toads, 177
Tools, human, 261–3
Tortoises, 206
Tracheophyta, 28, 149; early history of, 149–51; mode of life of, 149–50
Tracks, 42

Trails, 42
Triassic, 71; changes in life of, 189–90, 221, 279, 283; climate of, 221; life of, 173, 181–2, 189–90, 191–8, 221–7, 232, 270–76, 299
Triceratops, 203
Triconodonts, 230
Trilobites, 115–18, 121–4; history of, 117–22; mode of life of, 122–4
Tunicates, larval, as possible ancestors of vertebrates, 160
Turtles, 206
Tyrannosaurus, 199

Uintatheres, 235–6
Ungulates, genera of, 235; history of, 235; modern forms of, 235
Urey, H. C., 58
Ursus arctos, etruscos, and *spelaeus*, 287
Ussher, Archbishop, 284

Variation, 20–21, 27–8
Vertebrates, ancestry of, 109, 111, 159–63; characters of, 158–9; divisions of, 158–9; preparation of fossils of, 45; oldest, in Middle Ordovician, 159; origins of, 109, 111, 158–63
Viruses, 26–7

Walcott, C. D., 118
Wallace, Alfred Russell, 277–8, 283–5
Walruses, 251
Watson, 300
Weiner, J. S., 54
Whales, 250
Wing structure, 214
Woolly mammoth, 246
Worms, 107–9

Xenophanes, 34–5

Zones, climatic, 32

MORE ABOUT PENGUINS
AND PELICANS

Penguinews, which appears every month, contains details of all the new books issued by Penguins as they are published. From time to time it is supplemented by *Penguins in Print*, which is a complete list of all titles available. (There are some five thousand of these.)

A specimen copy of *Penguinews* will be sent to you free on request. For a year's issues (including the complete lists) please send 50p if you live in the British Isles, or 75p if you live elsewhere. Just write to Dept EP, Penguin Books Ltd, Harmondsworth, Middlesex, enclosing a cheque or postal order, and your name will be added to the mailing list.

In the U.S.A.: For a complete list of books available from Penguin in the United States write to Dept CS, Penguin Books Inc., 7110 Ambassador Road, Baltimore, Maryland 21207.

In Canada: For a complete list of books available from Penguin in Canada write to Penguin Books Canada Ltd, 41 Steelcase Road West, Markham, Ontario.

SCIENCE AND SOCIETY

Hilary Rose and Steven Rose

'So important that no one can afford to neglect it. Who is directing research, technological development, industry, and education toward the common good? The answer, as the complex arguments in this book demonstrate, is no one' – *Guardian*

In a study which invites comparison with J. D. Bernal's *The Social Function of Science* (published before the bomb or the cracking of the genetic code), a biochemist and a sociologist attack the notion that science, like fate, is 'an unpredictable act of gods in white coats'. Since it is the product of certain men in certain societies, it can be controlled. In their opening chapters they recount the history of science in its relations with society from the founding of the Royal Society to the post-war records of Conservative and Labour governments. For comparison they add chapters on the position in America, Russia, and other countries, and on the functions of international bodies (from Pugwash to UNESCO).

This book is recommended by *New Scientist* as 'a helpful starting-point ... for the students of the forthcoming Open University' in approaching the question of how science can be effectively harnessed for the good of all people.

DRUGS

Peter Laurie

What are the known facts about the 'dangerous' drugs? What actual harm, mental or physical, do they cause? Which of them are addictive, and how many addicts are there?

Peter Laurie has talked with doctors, policemen, addicts, and others intimately involved with this problem. He has tried some of the drugs himself and closely studied the medical literature (including little-known reports of American research).

The result of his inquiries into the pharmacological uses and social effects of drugs today appears in this book.

Originally published as a Penguin Special which went through five printings, *Drugs* was the first objective study to offer all the major medical, psychological and social facts about the subject to a public which is too often fed with alarmist and sensational reports. For this second edition in Pelicans Peter Laurie has added fresh information and statistics concerning English users of drugs and noted changes in the law.

MAN, MEDICINE AND ENVIRONMENT

René Dubos

In this book Dr Dubos, an editor of the *Journal of Experimental Medicine*, examines the environmental forces affecting the history of social groups from the precursors of *homo sapiens* to man today. Considering characteristics unique to humanity, he states that 'Man can function well only when his external environment is in tune with the needs he has inherited from his evolutionary, experiential, and social past, and with his aspirations for the future.' As man acquires much of his personality through responses to environment, Dr Dubos discusses the complex interrelations that govern life today, and the effects of environment on the health of primitive and modern man. In non-technical language he surveys the control of life, biomedical philosophies and the possibilities of a science of man. Precisely because they are concerned with various aspects of humanity, Dr Dubos believes that 'the biomedical sciences in their highest form are potentially the richest expression of science'.

Not for sale in the U.S.A. or Canada

GENES, DREAMS AND REALITIES

Macfarlane Burnet

'I cannot avoid the conclusion that we have reached the stage when little further advance can be expected from laboratory science in the handling of the "intrinsic" types of disability and disease.'

In this account of molecular biology and its potential contribution to human welfare, Sir Macfarlane Burnet administers a corrective to Utopian visions of a future in which age and disease have been conquered. And in addition he succeeds in making the problems of medical science intelligible to the layman.

Some will find his conclusions realistic, others will find them pessimistic; but his estimate of the possibilities of 'genetic engineering' and its application in the fields of human ageing, cancer research, mental disease and population control is made with all the authority of a scientist whose original work won a Nobel Prize.

'Lucid and authoritative and a pleasure to read' – *Economist*

'A rare feat: this book is written in such a way that a layman can understand it and a medical scientist enjoy it' – *Practitioner*

Not for sale in the U.S.A.